T0191498

METHODS IN MOLECULAR BIOLOGY

Series Editor
John M. Walker
School of Life and Medical Sciences
University of Hertfordshire
Hatfield, Hertfordshire, AL10 9AB, UK

For further volumes:
http://www.springer.com/series/7651

Nucleic Acid Aptamers

Selection, Characterization, and Application

Edited by

Günter Mayer

University of Bonn, Bonn, Germany

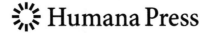 Humana Press

Editor
Günter Mayer
University of Bonn
Bonn, Germany

ISSN 1064-3745   ISSN 1940-6029   (electronic)
Methods in Molecular Biology
ISBN 978-1-4939-4934-2   ISBN 978-1-4939-3197-2   (eBook)
DOI 10.1007/978-1-4939-3197-2

Printed on acid-free paper

Humana Press is a brand of Springer
Springer Science+Business Media LLC New York is part of Springer Science+Business Media (www.springer.com)

Preface

Aptamers are short, single chained nucleic acids that fold into a well-defined three-dimensional shape, upon which a target molecule is recognized by high affinity and specificity. Since their very first description in 1990, a plethora of aptamers have been developed, characterized, and applied. Aptamers that recognize small molecules, proteins, and even targets on cell surfaces have been selected and applied for different purposes. Within the last decade, aptamers have become more and more popular, and their sophisticated biophysical properties together with their ability to be easily modified and, thus, adapted to various regimens make them a very promising class of compounds.

This book intends to provide protocol references covering recent developments in the aptamer field. It is subdivided into three parts reflecting the aptamer generation process, namely the selection, characterization, and finally the application of aptamers.

The selection part covers methods to isolate aptamers recognizing small molecules (Chapter 1) or to target molecules presented or embedded on cell surfaces (Chapters 2 and 3). In Chapter 4, the implementation of novel, noncanonical base pairs for generating new types of aptamers is described. A protocol for Capillary Electrophoresis-based selection schemes (CE-SELEX), an emerging approach to allow the generation of aptamers for non-tagged and non-immobilized target molecules, is given in Chapter 5. The first part of this volume of the *Methods in Molecular Biology* series is rounded out by two consecutive chapters dealing with Next Generation Sequencing for the thorough analysis of individual selection experiments (Chapters 6 and 7).

The second part covers a series of analytical methods to assess biophysical properties of aptamer–target interactions. This part contains protocols describing MicroScale Thermophoresis (Chapter 8), isothermal titration calorimetry (Chapter 9), and flow cytometry (Chapter 10) as means to measure dissociation constants and to address specificity of aptamers in regard to cognate ligand recognition. Chapter 11 covers methods to perform in vivo imaging of tissue-specific aptamers, whereas Chapter 12 provides an approach toward generating co-crystals of aptamer–target complexes for X-ray crystallography. With this, the second part of the volume is concluded leading into the third part, which covers various applications of aptamers.

In this way, Chapters 13–15 provide protocols for employing aptamers as detection modules in various assay formats, such as voltammetric assays, their simultaneous exploitation as recognition elements and PCR templates (Apta-PCR) and as capture-modules in the so-called enzyme-capture assays. The following chapters cover protocols for cellular applications of aptamers, e.g., their compatibility with histopathology investigations (Chapter 16) and their superior properties as imaging tools in super-resolution microscopy (Chapter 17). In Chapter 18, a protocol for the synthesis of liposomes decorated with cell-specific aptamers is described, which enables the application of aptamers as a targeting reagent for delivery of siRNAs into distinct cells. The last chapter provides means for generating genetic switches based on aptamer-responsive ribozymes (Chapter 19).

I am thankful to all contributors who made this book possible. I believe that this protocol collection provides a state-of-the-art summary of recent developments in the aptamer field and will be a helpful resource for scientists in the life sciences working with aptamers as tools to elucidate biological systems.

Bonn, Germany   *Günter Mayer*

Contents

PART III   APPLICATION

Contributors

PHILIPP BAASKE • *NanoTemper Technologies GmbH, Munich, Germany*

VINÍCIUS BASSANEZE • *Laboratory of Genetics and Molecular Cardiology/LIM 13, Heart Institute (InCor), University of São Paulo Medical School, São Paulo, Brazil*

TOBIAS BECHER • *Institute for Experimental Haematology and Transfusion Medicine, University of Bonn Medical Centre, Bonn, Germany*

KATHARINA BERG • *Chemistry Department, Institute for Biochemistry and Molecular Biology, MIN-Faculty, Hamburg University, Hamburg, Germany*

MICHAEL BLANK • *AptaIT GmbH, Munich, Germany*

DENNIS BREITSPRECHER • *NanoTemper Technologies GmbH, Munich, Germany*

SILVIA CATUOGNO • *Istituto per l'Endocrinologia e l'Oncologia Sperimentale del CNR "G. Salvatore", Naples, Italy*

GULSAH CONGUR • *Faculty of Pharmacy, Analytical Chemistry Department, Ege University, Bornov, Izmir, Turkey*

WEI DUAN • *School of Medicine, Deakin University, Waurn Ponds, VIC, Australia*

FRÉDÉRIC DUCONGÉ • *CEA, I2BM, Molecular Imaging Research Center (MIRCen), Fontenay-aux-Roses, France; INSERM U1023, Paris, France; Université Paris Sud, Paris, France*

STEFAN DUHR • *NanoTemper Technologies GmbH, Munich, Germany*

ECE EKSIN • *Faculty of Pharmacy, Analytical Chemistry Department, Ege University, Bornov, Izmir, Turkey*

ARZUM ERDEM • *Faculty of Pharmacy, Analytical Chemistry Department, Ege University, Bornov, Izmir, Turkey*

CARLA LUCIA ESPOSITO • *Istituto per l'Endocrinologia e l'Oncologia Sperimentale del CNR "G. Salvatore", Naples, Italy*

MICHELE FELLETTI • *Department of Chemistry and Konstanz Research School Chemical Biology, University of Konstanz, Konstanz, Germany*

VITTORIO DE FRANCISCIS • *Istituto per l'Endocrinologia e l'Oncologia Sperimentale del CNR "G. Salvatore", Naples, Italy*

KARINE GOMBERT • *INSERM U1023, Paris, France; Université Paris Sud, Paris, France; CEA, I2BM, Service Hospitalier Frédéric Joliot (SHFJ) Paris, France*

MARIA ANGELA GOMES DE CASTRO • *Department of Neuro- and Sensory Physiology, Cluster of Excellence Nanoscale Microscopy and Molecular Physiology of the Brain, University of Göttingen Medical Center, Göttingen, Germany*

ULRICH HAHN • *Chemistry Department, Institute for Biochemistry and Molecular Biology, MIN-Faculty, Hamburg University, Hamburg, Germany*

NASIM SHAHIDI HAMEDANI • *Institute for Experimental Haematology and Transfusion Medicine, University of Bonn Medical Centre, Bonn, Germany*

JÖRG S. HARTIG • *Department of Chemistry and Konstanz Research School Chemical Biology, University of Konstanz, Konstanz, Germany*

ICHIRO HIRAO • *Center for Life Science Technologies (CLST), RIKEN, Yokohama, Kanagawa, Japan; TagCyx Biotechnologies, Yokohama, Kanagawa, Japan*

MICHIKO KIMOTO • *Center for Life Science Technologies (CLST), RIKEN, Yokohama, Kanagawa, Japan; TagCyx Biotechnologies, Yokohama, Kanagawa, Japan; PRESTO, JST, Saitama, Japan*

BENEDIKT KLAUSER • *Department of Chemistry and Konstanz Research School Chemical Biology, University of Konstanz, Konstanz, Germany*

JOSÉ E. KRIEGER • *Laboratory of Genetics and Molecular Cardiology/LIM 13, Heart Institute (InCor), University of São Paulo Medical School, São Paulo, Brazil*

BENOIT LELANDAIS • *CEA, I2BM, Molecular Imaging Research Center (MIRCen), Fontenay-aux-Roses, France; INSERM U1023, Paris, France; Université Paris Sud, Paris, France*

TERESA MAIRAL LERGA • *Department of Chemical Engineering, Universitat Rovira I Virgili, Tarragona, Spain*

MATTHEW LEVY • *Department of Biochemistry, Michael F. Price Center for Genetic and Translational Medicine, Albert Einstein College of Medicine, Bronx, NY, USA*

YI LONG • *Institute of Biochemistry and Molecular Biology, University of Southern Denmark, Odense, Denmark; LIMES Institute, University of Bonn, Bonn, Germany*

LI LV • *Department of Pathology, Second Affiliated Hospital of Dalian Medical University, Dalian, China*

EILEEN MAGBANUA • *Chemistry Department, Institute for Biochemistry and Molecular Biology, MIN-Faculty, Hamburg University, Hamburg, Germany*

KEN-ICHIRO MATSUNAGA • *Center for Life Science Technologies (CLST), RIKEN, Yokohama, Kanagawa, Japan; TagCyx Biotechnologies, Yokohama, Kanagawa, Japan*

GÜNTER MAYER • *Life & Medical Sciences Institute (LIMES), University of Bonn, Bonn, Germany*

JENS MÜLLER • *Institute for Experimental Haematology and Transfusion Medicine, University of Bonn Medical Centre, Bonn, Germany*

ISIS C. NASCIMENTO • *Departamento de Bioquímica, Instituto de Química, Universidade de São Paulo, São Paulo, Brazil*

ARTHUR A. NERY • *Departamento de Bioquímica, Instituto de Química, Universidade de São Paulo, São Paulo, Brazil*

CIARA K. O'SULLIVAN • *Department of Chemical Engineering, Universitat Rovira I Virgili, Tarragona, Spain; Institució Catalana de Recerca I Estudis Avançats, Barcelona, Spain*

LARS FOLKE OLSEN • *Institute of Biochemistry and Molecular Biology, University of Southern Denmark, Odense, Denmark*

FELIPE OPAZO • *Department of Neuro- and Sensory Physiology, Cluster of Excellence Nanoscale Microscopy and Molecular Physiology of the Brain, University of Göttingen Medical Center, Göttingen, Germany*

VELI CENGIZ ÖZALP • *Department of Medical Biology, School of Medicine, Istanbul Kemerburgaz University, Istanbul, Turkey*

FRANZISKA PFEIFFER • *LIMES Institute, University of Bonn, Bonn, Germany*

ALESSANDRO PINTO • *Department of Chemical Engineering, Universitat Rovira I Virgili, Tarragona, Spain; Department of Bioengineering, Rice University, Houston, TX, USA*

PEDRO NADAL POLO • *Department of Chemical Engineering, Universitat Rovira I Virgili, Tarragona, Spain; Center for Omic Sciences (COS), Universitat Rovira I Virgili, Reus, Spain*

BERND PÖTZSCH • *Institute for Experimental Haematology and Transfusion Medicine, University of Bonn Medical Centre, Bonn, Germany*

NAM NGUYEN QUANG • *CEA, I2BM, Molecular Imaging Research Center (MIRCen), Fontenay-aux-Roses, France; INSERM U1023, Université Paris Sud, Paris, France*

BURKHARD RAMMNER • *Department of Neuro- and Sensory Physiology, Cluster of Excellence Nanoscale Microscopy and Molecular Physiology of the Brain, University of Göttingen Medical Center, Göttingen, Germany*

MIRIAM JAUEST RUBIO • *Department of Chemical Engineering, Universitat Rovira I Virgili, Tarragona, Spain*

NINA SCHLINCK • *NanoTemper Technologies GmbH, Munich, Germany*

TINE DAA SCHRØDER • *Institute of Biochemistry and Molecular Biology, University of Southern Denmark, Odense, Denmark; Lundbeckfonden Center of Excellence NanoCAN, University of Southern Denmark, Odense, Denmark*

THOMAS SCHUBERT • *2bind GmbH, Regensburg, Germany*

SARAH SHIGDAR • *School of Medicine, Deakin University, Waurn Ponds, VIC, Australia*

BEATRIX SUESS • *Department of Biology, Technical University Darmstadt, Darmstadt, Germany*

MARKETA SVOBODOVA • *Department of Chemical Engineering, Universitat Rovira I Virgili, Tarragona, Spain*

JOHN J.G. TESMER • *Departments of Pharmacology and Biological Chemistry, University of Michigan, Ann Arbor, MI, USA*

IOANNA THÉODOROU • *CEA, I2BM, Molecular Imaging Research Center (MIRCen), Fontenay-aux-Roses, France; INSERM U1023, Université Paris Sud, Paris, France*

BENOIT THÉZÉ • *INSERM U1023, Paris, France; Université Paris Sud, Paris, France; CEA, I2BM, Service Hospitalier Frédéric Joliot(SHFJ), Paris, France*

FABIAN TOLLE • *Life & Medical Sciences Institute (LIMES), University of Bonn, Bonn, Germany*

HENNING ULRICH • *Departamento de Bioquímica, Instituto de Química, Universidade de São Paulo, São Paulo, Brazil*

MARC VOGEL • *Department of Biology, Technical University Darmstadt, Darmstadt, Germany*

LIFEN WANG • *Department of Pathology, Second Affiliated Hospital of Dalian Medical University, Dalian, China*

SAMANTHA E. WILNER • *Department of Biochemistry, Michael F. Price Center for Genetic and Translational Medicine, Albert Einstein College of Medicine, Bronx, NY, USA*

DAVID WITTE • *NanoTemper Technologies GmbH, Munich, Germany*

Part I

Selection

Chapter 1

Selection of Aptamers for Metabolite Sensing and Construction of Optical Nanosensors

Yi Long, Franziska Pfeiffer, Günter Mayer, Tine Daa Schrøder, Veli Cengiz Özalp, and Lars Folke Olsen

Abstract

Optical nanosensors are based on particles with diameters from 20 to 200 nm containing sensory elements. The latter are comprised of one or more signaling molecules and one or more references, which allow measurements to be ratiometric and hence independent on the amount of sensor. The signaling molecules may range from simple ion-binding fluorophores, e.g., pH-sensitive dyes, to complex biochemical assays. Aptamers are ideal for use in nanosensors because they are relatively easy to modify chemically and hence to transform into signaling molecules, and their binding affinities may be fine-tuned to a desired measuring range in the selection process. Here we first describe the selection of metabolite binding aptamers, how they are transformed into signaling molecules using a molecular beacon construct and then how they are inserted into nanoparticles. Finally, we briefly describe how the sensors are calibrated before inserted into cells to measure metabolite concentration in real time. As examples we present aptamers binding to key metabolites in cells: ATP and fructose 1, 6-bisphosphate (FBP).

Key words Nanosensor, Aptamers, Selection, 6-Bisphosphate, Fructose 1, ATP, Calibration

1 Introduction

Metabolism in living cells is highly dynamic and variable. The cell must be able to respond to external stimuli within seconds and sometimes oscillations in physiologic concentrations of intracellular metabolites with periods ranging from seconds to minutes to hours are observed [1]. In order to understand metabolic processes and eventually to simulate them by mathematical models, it is very important that we can measure metabolites in real time in intact cells. This is not an easy task as most metabolites are generally spectroscopically "silent." There are only a few exceptions to this rule, such as NAD(P)H and flavines, which are fluorescent. To measure other intracellular metabolites we usually have to quench

Günter Mayer (ed.), *Nucleic Acid Aptamers: Selection, Characterization, and Application*, Methods in Molecular Biology, vol. 1380, DOI 10.1007/978-1-4939-3197-2_1, © Springer Science+Business Media New York 2016

cells and extract their content, which is subsequently measured using off-line methods [2]. Therefore, measurements of metabolites are usually performed at long discrete time intervals using the assumption that metabolism is in a stationary state. In order to increase our knowledge about metabolism, there are increased efforts to develop sensors, which can be inserted into cells in a noninvasive, or at least minimal invasive way, to measure intracellular metabolites in real time. Some of these sensors are genetically engineered fluorescent or Förster resonance energy transfer (FRET)-based proteins [3–5], while others are based on nanoparticles that encapsulate sensing elements [6, 7]. For the latter aptamers come in handy because (1) the SELEX [8, 9] technique, in principle, allows for the construction of aptamers to essentially any molecule, including low-molecular-weight metabolites, and with a predetermined affinity, and (2) aptamers can be modified chemically, which makes it relatively easy to transform them into signaling molecules [10, 11].

Here we will describe the process of selecting an aptamer to an important metabolite in cells: Fructose 1, 6-bisphosphate (FBP). FBP is a key intermediate in glycolysis, and it also serves as a regulator of other metabolic processes. In addition to the selection of the FBP aptamer, we will describe how aptamers may be converted into signaling molecules that are inserted into polyacrylamide particles to form optical nanosensors. Finally, we briefly describe how these sensors are calibrated before they are inserted into intact cells to measure intracellular metabolites in real time [12, 13].

2   Materials

2.1   Selection Matrix

2.1.1   Negative Selection Matrix

1. Sepharose 4B (Sigma).
2. Empty polypropylene spin columns, 1.2 mL bed volume (Biorad).
3. 1.5 mL Eppendorf tubes.
4. 5× Cytosol buffer: 0.03 M KH_2PO_4, 0.07 M K_2HPO_4, 0.7 M KCl, 0.05 M NaCl, 27.5 % glucose. Adjust to pH 6.88. Store at –20 °C (*see* **Note 1**).

2.1.2   Functionalization of Sepharose with D-Fructose 1, 6-Bisphosphate

1. Sepharose 4B (Sigma).
2. D-fructose 1, 6-bisphosphate (FBP).
3. 1, 4-Butanediol diglycidyl ether.
4. 0.5 M Ethanolamine dissolved in water.
5. 0.6 M NaOH with 0.6 % $NaBH_4$ in water.
6. 0.5 M Na_2CO_3 in water.

7. Filtering setup (one Büchner flask, one glass filter funnel).

8. A shaker with heating.

2.1.3 Positive Selection Matrix

1. FBP-labeled Sepharose 4B.

2. Empty polypropylene spin columns, 1.2 mL bed volume (Biorad).

3. 1.5 mL Eppendorf tubes.

4. 1× Selection buffer: Dilute 5× cytosol buffer to 1× cytosol buffer by mixing with water and 100 mM $MgCl_2$ to obtain 1× cytosol buffer with 5 mM $MgCl_2$.

2.2 Selection of FBP Aptamer

2.2.1 SELEX Round 1–3

1. Sepharose 4B (Sigma).

2. FBP-labeled Sepharose 4B.

3. Thermomixer.

4. 5× Cytosol buffer.

5. 100 mM $MgCl_2$ in water.

6. RNA library 5′-GGGAGGACGAUGCGG-N40-CAGACGAC UCGCUGAGGAUCCGAGA-3′ [14].

7. 1× Selection buffer.

8. 5 mM EDTA, pH 8.0 in water.

9. 3 M sodium acetate, pH 5.4.

10. 99.8 % ethanol.

11. 10 mg/mL glycogen in water.

12. 70 % ethanol.

2.2.2 Reverse Transcription Polymerase Chain Reaction (RT-PCR)

1. 5× Colorless GoTaq Flexi Buffer (Promega, cat# M890A).

2. 5× first-strand buffer (Invitrogen).

3. 100 mM dithiothreitol (DTT) in water.

4. 100 mM $MgCl_2$.

5. 25 mM (each) dNTPs.

6. 100 µM forward primer: 5′-GGGGGAATTCTAATACGACT CACTATAGGGAGGACGATGCGG-3s.

7. 100 µM reverse primer: 5′-TCTCGGATCCTCAGCGAGT CGTC-3′.

8. Superscript II Reverse Transcriptase (200 U/µL) (Invitrogen, cat# 18064).

9. GoTaqFlexiDNA polymerase (5 U/µL) (Promega, cat# M8305).

10. Agarose, electrophoresis grade.

11. 10× TBE-buffer: 89 mM Tris, pH 8.0, 89 mM boric acid, 2 mM EDTA. Dilute 1:10 with water.

12. 10 mg/mL Ethidium bromide (*see* **Note 2**).

13. Agarose gel-loading buffer: 50 % glycerol, 50 mM Tris–HCl, pH 8.0, 50 mM EDTA, optionally add one spatula tip of bromophenol blue and xylene cyanol.

2.2.3   Transcription

1. 5× Transcription buffer: 200 mM Tris–HCl, pH 7.9.

2. 100 mM DTT in water.

3. 25 mM (each) NTP-Mix.

4. 100 mM MgCl$_2$ in water.

5. 40 U/μL Recombinant RNasin Ribonuclease Inhibitor (Promega, cat# N2515/N2511).

6. 30 U/μL T7 RNA Polymerase.

7. Agarose, electrophoresis grade.

8. 10× TBE-buffer: 89 mM Tris, pH 8.0, 89 mM boric acid, 2 mM EDTA. Dilute 1:10 with water.

9. 10 mg/mL Ethidium bromide.

10. Agarose gel-loading buffer: 50 % glycerol, 50 mM Tris–HCl, pH 8.0, 50 mM EDTA, optionally add one spatula tip of bromophenol blue and xylene cyanol.

2.2.4   SELEX Round 4-×

1. FBP-labeled Sepharose 4B.

2. 5× Cytosol buffer: 0.03 M KH$_2$PO$_4$, 0.07 M K$_2$HPO$_4$, 0.7 M KCl, 0.05 M NaCl, 27.5 % glucose, pH 6.88.

3. 100 mM MgCl$_2$.

4. 1× Selection buffer: 1× Cytosol buffer with 5 mM MgCl$_2$.

5. 1× Elution buffer: 1× Cytosol buffer with 5 mM MgCl$_2$ and 5 mM FBP.

6. 3 M Sodium acetate, pH 5.4.

7. 99.8 % Ethanol.

8. 10 mg/mL glycogen in water.

2.3   5′-End Labeling of RNA Molecules

1. 5× Transcription buffer: 200 mM Tris–HCl, pH 7.9.

2. 100 mM DTT.

2.3.1   Large-Scale Transcription

3. 25 mM (each) NTP-Mix.

4. 100 mM MgCl$_2$.

5. 40 U/μL Recombinant RNasin Ribonuclease Inhibitor (Promega, cat# N2515/N2511).

6. 30 U/μL T7 RNA Polymerase stored in aliquots at –20 °C in 20 mM sodium phosphate, pH 7.7, 1 mM DTT, 1 mM EDTA, 100 mM NaCl and 50 % glycerol.

7. Inorganic pyrophosphatase (IPP) (2 U/μL) (Roche).

8. Nano Quant (Tecan nanoquant infinite M200).

2.3.2 Dephosphorylation

1. 10× Calf intestine alkaline phosphatase (CIAP) buffer (Promega).
2. 10 mg/mL Bovine serum albumin (BSA, RNase-free grade) stored at 4 °C in water.
3. CIAP enzyme (Promega).
4. 40 U/μL Recombinant RNasin Ribonuclease Inhibitor (Promega, cat# N2515/N2511).
5. 500 mM EDTA, pH 8.0 in water.
6. Phenol and chloroform.
7. 3 M Sodium acetate, pH 5.4 in water.
8. 99.8 % Ethanol.
9. 10 mg/mL Glycogen in water.
10. 70 % Ethanol.

2.3.3 5′-Radioactive Phosphorylation

1. 10 U/μl T4 Polynucleotide-Kinase (T4 PNK).
2. 10× T4 PNK buffer (New England Biolabs).
3. γ-[^{32}P]-ATP (*see* **Note 3**).
4. Illustra MicroSpin G-25 columns.

2.4 Polyacrylamide Gel Electrophoresis

1. 70 % Ethanol.
2. Plates, spacers, clamps, comb.
3. Concentrated gel solution, 25 % Bis-/Acrylamide in 8.3 M Urea (*see* **Note 4**).
4. 8.3 M Urea in water.
5. 10× TBE-buffer: 89 mM Tris, pH 8.0, 89 mM boric acid, 2 mM EDTA.
6. 10 % Ammoniumperoxodisulphate solution (APS). Store at 4 °C.
7. *N*,*N*,*N*′,*N*′-Tetramethylethylenediamine (TEMED). Store at 4 °C.
8. 4× Polyacrylamide gel electrophoresis (PAGE)-Gel loading buffer: 9 M Urea, 50 mM EDTA, pH 8.0.
9. Bromophenol blue and xylene cyanol.

2.5 PAGE Gel Purification

1. Phosphorimager (Fujifilm FLA-3000).
2. BAS cassette 2023 and BAS MS imaging plates (Fujifilm FLA-3000).
3. Sterile scalpel.
4. 0.3 M sodium acetate, pH 5.4.
5. Thermomixer.
6. Syringe.
7. Glass wool (RNase-free grade).

2.6 Binding Assay

1. Sepharose 4B (Sigma).
2. FBP-labeled Sepharose 4B.
3. Empty polypropylene spin columns, 1.2 mL bed volume (Biorad).
4. 1.5 mL Reaction vials.
5. Purified 5′-end radioactively labeled RNA.
6. Thermomixer.
7. Liquid scintillation counter.
8. 5× Cytosol buffer: 0.03 M KH_2PO_4, 0.07 M K_2HPO_4, 0.7 M KCl, 0.05 M NaCl, 27.5 % glucose, pH 6.88.
9. 100 mM $MgCl_2$.
10. 1× Selection buffer: 1× Cytosol buffer with 5 mM $MgCl_2$.
11. 1× Elution buffer: 1× Cytosol buffer with 5 mM $MgCl_2$ and 5 mM FBP.

2.7 Synthesis of Aptamer-Nanosensors

1. ATP aptamer: 5′-GTAGTAAGAACTAAAGTAAAAAAAAA ATTAAAGTAGCCACGCTT-3′.
2. Aptamer switch probe: 5′ (AF488)-GTAGTAAGAACTAAAG T A A A A A A A A A A T T A A A G T A G C C A C G C T T - [(C18spacer)×6]-TTACTAC-(BHQ1) 3′ (*see* **Note 5**) (VBC Biotech, Vienna).
3. Dioctyl sulfosuccinate.
4. Polyethylene glycol dodecyl ether (Brij 30).
5. Acrylamide solution of 38.8%T, 1.35 g of acrylamide, 0.4 g of N,N-methylenebisacrylamide.
6. 10 mg/mL Texas Red Dextran solution, MW of dextran: 10,000 g/mol.
7. 10 % Sodium bisulfite: 50 mg Na-bisulfite dissolved in 0.45 mL 10 mM Na-phosphate buffer, pH 7.2.
8. Three-necked 100 mL round-bottomed flask.
9. Sonication bath.
10. Argon gas.
11. Amicon ultrafiltration cell model 2800 (Millipore Corp., Bedford, USA).
12. 100 kDa filter (Ultrafiltration Membrane YM100).
13. 0.025 μm nitrocellulose filter membrane (Millipore filter type VS 0.025 μm).

2.8 Calibration of Nanosensors

1. Nanosensors containing aptamer switch probe and reference fluorophore.
2. 1× Cytosol buffer with 5 mM $MgCl_2$.
3. 500 mM ADP in 1× cytosol buffer, pH 7.0.

4. 500 mM Phosphoenolpyruvate in water.

5. 1 U/µL Rabbit pyruvate kinase.

6. 1 U/µL Yeast hexokinase.

3 Methods

3.1 Functionalization of Sepharose with D-Fructose-1, 6-Bisphosphate

1. Wash 10 g sepharose 4B with 500 mL H_2O on a filtering setup using suction. Remove as much water as possible by suction.

2. Resuspend the sepharose in 10 mL 0.6 M NaOH containing 0.6 % $NaBH_4$. Subsequently add 5 mL 1, 4-butanediol diglycidyl ether (liquid ~ 27.2 mmol).

3. Leave the suspension to react overnight on a shaker (150 rpm) at room temperature.

4. Wash the sepharose with 1.5 L H_2O. Remove as much water as possible by suction.

5. Resuspend the activated sepharose in 50 mL 0.5 M Na_2CO_3, and then add 4.5 g FBP (8.2 mmol). Shake the mixture overnight at 50 °C.

6. Wash the sepharose with 1.5 L H_2O. At the end, remove as much water as possible by suction.

7. Resuspend the sepharose in 50 mL 0.5 M ethanolamine and shake it for 1 h at room temperature.

8. Wash the labeled sepharose with 1.5 L H_2O and resuspend it in 10 mL H_2O. Store at 4 °C.

3.2 Selection of FBP Aptamer

3.2.1 SELEX Rounds 1–3

1. The selection matrix is prepared by loading 100 µL of FBP-labeled sepharose for positive selection and sepharose 4B for negative selection into empty polypropylene spin columns and washing them with 600 µL H_2O.

2. In SELEX round 1: Mix 1 nmol of the RNA library with 60 µL 5× cytosol buffer, 15 µL 100 mM $MgCl_2$, and fill H_2O to achieve 300 µL total volume. In SELEX rounds 2–3: Mix 50 µL transcription product from the previous selection cycle with 60 µL 5× cytosol buffer, 7.5 µL 100 mM $MgCl_2$ and fill with H_2O to achieve 300 µL total volume.

3. Put the matrix for negative selection into a thermomixer, which is set to 25 °C, load the sample solution (*see* **step 2**) onto the negative matrix, let it pass by gravity, and collect the flow through.

4. Apply the flow through from **step 3** to the FBP-labeled selection matrix. Discard the flow through.

5. Wash the column twice with 600 µL 1× selection buffer.

6. Add 300 µL of 5 mM EDTA to the column to elute the bound RNA and collect the eluted fraction.

7. Add 30 μL of 3 M sodium acetate, 990 μL of 99.8 % ethanol, and 1 μL of 10 mg/mL glycogen to the eluted fraction.

8. Cool the mixture at −80 °C for 30 min and subsequently centrifuge at $20,000 \times g$ for 20 min.

9. Discard the supernatant and wash the pellet with 70 % ethanol.

10. Dry the pellet and resuspend it in 50 μL of H_2O.

3.2.2 Reverse Transcription Polymerase Chain Reaction

1. Prepare a reverse transcription polymerase chain reaction (RT-PCR) master mix as seen in Table 1.

2. Add 50 μL H_2O to one tube of the RT-PCR master mix as a negative control and 50 μL of the purified eluate fraction (*see* previous section, **steps 6–10**) to a second tube.

3. Incubate the two samples at 65 °C for 5 min.

4. Immediately place the samples at 4 °C.

5. Add 1 μL of Superscript II Reverse Transcriptase (200 U/μL) and 1 μL of GoTaq Flexi DNA polymerase (5 U/μL) to the samples.

6. Incubate for 10 min at 54 °C.

7. Place the sample in the thermocycler with the following settings: 95 °C for 30 s, 60 °C for 30 s, and 72 °C for 30 s (*see* **Note 6**).

8. Prepare a 2.5 % (w/v) agarose gel in 1× TBE-buffer and boil it until the agarose is completely dissolved.

9. Add 0.1 μL/mL of 10 mg/mL ethidium bromide and mix gently.

10. Pour the gel into a gel-casting chamber, insert the comb, and let it solidify.

Table 1
Protocol for setting up the RT-PCR reaction

Reagent	Volume (μL)/reaction	Final concentration
5× Colorless GoTaq Flexi Buffer	20	1×
5× first-strand buffer	4	0.2×
100 mM DTT	2	2 mM
100 μM Primer fw Sul 1	1	1 μM
100 μM Primer rv Sul 1	1	1 μM
100 mM $MgCl_2$	1.5	1.5 mM
25 mM (*each*) dNTPs	1.2	300 μM
H_2O	17.3	
Total	48	

11. Mix 2 μL PCR product with 2 μL agarose-loading buffer and load it onto the gel. Also load a suitable DNA-ladder (e.g., 100 bp ladder).

12. Run the gel at 160 V with 1× TBE as running buffer for 20 min and visualize the DNA bands with a state-of-the-art gel documentation system. Figure 1 shows an exemplary agarose gel image of PCR products obtained from different selection cycles.

3.2.3 Transcription

1. Prepare the transcription master mix as seen in Table 2.

2. For one reaction add 10 μL RT-PCR product to 85.7 μL of transcription master mix. Subsequently, add 1 μL RNasin (40 U/μL) and 3.3 μL T7 RNA Polymerase (30 U/μL).

3. Perform the transcription reaction for 20 min at 37 °C.

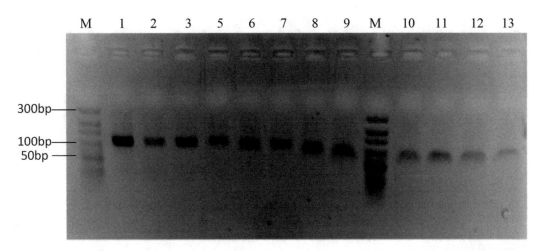

Fig. 1 RT-PCR product of different selection cycles. *M*: DNA size marker; *numbers 1–13* indicate the SELEX cycle

Table 2
Protocol for setting up the transcription reaction

Reagent	Volume (μL) for one reaction	Volume (μL) for 20 reactions	Final concentration
5× Transcription buffer	20	400	1×
100 mM DTT	5	100	5 mM
NTP-Mix (25 mM each)	10	200	2.5 mM each
100 mM MgCl$_2$	15	300	15 mM
H$_2$O	35.7	710	
Total volume	85.7	1710	

3.2.4 SELEX Round 4-x

1. Mix 50 μL transcription product from the previous selection cycle with 60 μL 5× cytosol buffer and 7.5 μL 100 mM MgCl$_2$ and add up to 300 μL with H$_2$O.

2. Apply the mixture with the transcription product from the previous round to the FBP-labeled sepharose and discard the flow through.

3. Wash the column twice with 600 μL 1× selection buffer.

4. Apply 300 μL of 1× elution buffer to the column to elute the bound RNA and collect the eluted fraction.

5. Add 30 μL of 3 M sodium acetate, 990 μL of 99.8 % ethanol, and 1 μL 10 mg/mL glycogen to the eluted fraction.

6. Freeze the mixture at −80 °C for 30 min; centrifuge it at 4 °C and 20,000 × g for 20 min.

7. Discard the supernatant, dry the pellet and resuspend it in 50 μL of H$_2$O.

3.3 5′-End Labeling of RNA Molecules

3.3.1 Large-Scale Transcription

1. Mix 200 μL 5× transcription buffer with 10 μL RNasin (40 U/μL), 2 μL IPP (2 U/μL), 30 μL T7 RNA Polymerase (30 U/μL), and 258 μL H$_2$O.

2. Add 100 μL RT-PCR product to the solution.

3. Incubate the reaction for 4 h at 37 °C.

4. Purify the RNA with a MicroSpin G-25 column.

5. Measure the RNA concentration on Nano Quant. Take at least 75 pmol RNA to do the dephosphorylation.

3.3.2 Dephosphorylation

1. Mix 5 μL 10× CIAP buffer, 5 μL 10 mg/mL BSA, 0.85 μL 20 U/μL CIAP, 0.5 μL 40 U/μL RNasin, and 75 pmol RNA and fill it up to 50 μL with H$_2$O.

2. Incubate at 37 °C for 15 min.

3. Add further 0.425 μL of 20 U/μL CIAP.

4. Incubate for an additional 15 min at 55 °C.

5. Add 0.5 μL 0.5 M EDTA, pH 8.0.

6. Incubate at 75 °C for 10 min.

7. Add 150 μL H$_2$O.

8. Phenol/chloroform extraction: Mix 200 μL dephosphorylated RNA with 200 μL of phenol and centrifuge at 14,000 × g for 10 min. Take the upper phase and mix it with 400 μL of chloroform; centrifuge at 14,000 × g for 10 min. Take the upper phase and add 1/10 volume of 3 M sodium acetate, pH 5.4, 3 volumes ethanol and 1 μL of 10 mg/mL glycogen.

9. Cool the mixture at −80 °C for 30 min and then centrifuge it at 4 °C and 20,000 × g for 20 min. Discard the supernatant.

10. Dry the pellet and resuspend it in 5 μL H$_2$O. Use 2 μL to run an agarose gel and 3 μL for the phosphorylation.

3.3.3 5′-Labeling with ³²P-ATP

1. Mix 3 μL dephosphorylated RNA, 2 μL 10× T4 PNK buffer, 1–2 μL γ-[³²P]-ATP (*see* **Note 7**), 2 μL 10 U/μL T4 PNK and fill up to 20 μL with H_2O.

2. Incubate for 30–45 min at 37 °C.

3. Purify the RNA with a MicroSpin G-25 column.

3.4 Polyacrylamide Gel Electrophoresis (PAGE)

1. Clean the glass plates with water and 70 % ethanol before use.

2. Prepare 30 mL of a 10 % PAGE-gel by mixing 12 mL 25 % Bis-/Acrylamide containing 8.3 M urea, 15 mL 8.3 M urea in water, 3 mL 8.3 M urea in 10× TBE, 240 μL 10 % APS, and 12 μL TEMED. Pour the mixture into the space between the two glass plates and insert the comb. Let it polymerize for at least 30 min.

3. Carefully remove the comb and the spacer at the bottom of the glass plates. Assemble the gel-running unit. Fill the lower chamber with running buffer and remove any air bubbles between gel and buffer. Afterwards, fill the upper chamber with buffer and rinse the wells with running buffer using a syringe.

4. Mix 50 μL radioactively labeled RNA sample from previous section **step 3**, with 17 μL 4× PAGE-gel loading buffer, heat it for 3 min at 80 °C and load onto the gel immediately.

5. Run the gel with 15 W/gel until the xylene cyanol band has migrated through 2/3 of the gel.

6. Remove the gel from the gel-running unit and the glass plates and wrap it in plastic foil.

7. Put the gel into an irradiation cassette and place an imaging plate on top, close the cassette and expose the screen for 10 min. Read the screen using a phosphorimager. Print the image at 100 % scale.

8. Place the printout underneath the gel, find the position of the sample, cut out the band with a sterile scalpel, and place the gel pieces in a 2 mL reaction vial. Crush the gel pieces with a blue pipet tip.

9. Add 0.3 M sodium acetate pH 5.4 to resuspend the crushed gel matrix.

10. Incubate the gel pieces at 65 °C at 1300 rpm and for 2 h on a thermomixer.

11. Filter the suspension through a syringe with glass wool and wash the gel pieces at least three times with 0.3 M sodium acetate. Collect the filtrate.

12. Add 3 volumes 99.8 % ethanol and 1 μL 10 mg/mL glycogen.

13. Leave at –20 °C overnight or cool for 30 min at –80 °C.

14. Centrifuge at $14,000 \times g$ and 4 °C for 20 min.

15. Discard the supernatant, dry the pellet, and dissolve it in 50 μL MilliQ H_2O.

3.5 Binding Assay

1. The selection matrix is prepared by loading 100 μL each of FBP labeled sepharose for positive selection and sepharose 4B for negative selection into a spin column and washing them with 600 μL H_2O each.

2. Mix 50 μL of radioactively labeled RNA (approx. 100,000 cpm), 60 μL 5× cytosol buffer, and 15 μL 100 mM $MgCl_2$ and fill it up to 300 μL with H_2O.

3. Put the sepharose 4B or FBP-labeled sepharose into a thermomixer at 25 °C, apply the radioactive RNA from above, and collect the flow through.

4. Apply 600 μL 1× selection buffer to the matrix, let it pass through by gravity, and collect the fraction. Repeat this procedure.

5. Apply 600 μL 1× elution buffer to the matrix, let it pass through, and collect this fraction.

6. Add H_2O to all samples to yield 1 mL final volume.

7. Measure the counts per minute by liquid scintillation counting. One representative result of a typical binding assay is shown in Fig. 2.

3.6 Synthesis of Aptamer-Nanoparticles

1. In a three-necked 100 mL round-bottomed flask purged with argon place 3.08 g dioctyl sulfosuccinate and 1.59 g Brij 30 and add 45 mL hexane.

2. Connect the flask to the argon, close it with stoppers, and leave it on the magnetic stirrer.

3. Deoxygenate by sonication for 1 h.

4. Prepare separately a solution of 38.8%T solution of acrylamide in a 7 mL vial purged with argon: 1.35 g Acrylamide, 0.4 g N,N-methylenebisacrylamide, 4.5 mL 10 mM Na-phosphate buffer, pH 7.2

5. Add 15 nmol of ATP aptamer probe and 15 μL Texas Red dextran (10 mg/mL) into the acrylamide solution and mix for 15 min.

6. When the dioctyl sulfosuccinate is dissolved move the flask to the sonication bath and remove the stopper in the small neck. Put also the monomer solution in the sonication bath without a cap. Both solutions are left for 1 h in the sonication bath. After sonication the round-bottomed flask is closed with a stopper and moved back to the magnetic stirrer.

7. Add 2.0 mL of acrylamide solution into hexane solution dropwise (*see* **Note 8**).

8. Stir solution for 15–20 min under argon such that the microemulsion can form.

9. Initiate polymerization by adding 50 μL of 10 % w/v solution of sodium bisulfite (*see* **Note 8**).

10. Allow the reaction to proceed for 3 h.

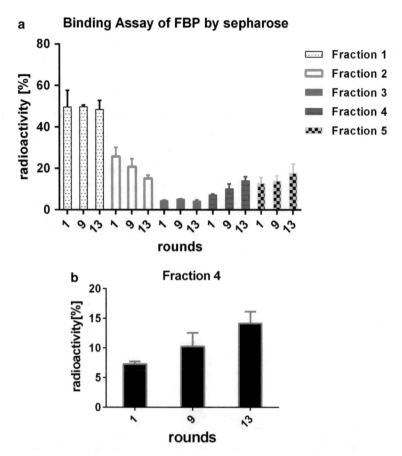

Fig. 2 Binding assay results for RNA product of different rounds. (**a**) Percentage of radioactivity in different fractions. Fraction 1: Eluate from different rounds of transcribed radioactively labeled RNA in 1× selection buffer. Fraction 2: The first wash flow-through by 1× selection buffer. Fraction 3: The second wash flow through by 1× selection buffer. Fraction 4: Eluted RNA by elution buffer. Fraction 5: Radioactivity left on sepharose. (**b**) Fraction 4: Eluted RNA from different selection rounds. The figure shows that, as the selection round goes up, the percentage of radioactivity for RNA from FBP-labeled sepharose also increases, which means the SELEX procedure enriched the amount of RNA that can bind to FBP

11. Remove hexane under vacuum in a rotation evaporator.

12. Add 100 mL 96 % ethanol to the nanosensor suspension.

13. Transfer the suspension to an Amicon ultrafilration cell with 100 kDa filter.

14. Wash 4 times with 100 mL of 96 % ethanol to remove unreacted monomers and surfactants. Do not let the filter cake run dry.

15. Resuspend washed particles in 50 mL of 96 % ethanol and filter through 0.025 μm nitrocellulose filter membrane.

16. Dry under vacuum and weigh (a typical synthesis yields about 400 mg of particles).

17. Keep dried particles at –20 °C until use.

3.7 Nanosensor Calibration

1. ATP-nanosensors are dissolved in 2 mL 1× cytosol buffer and 5 mM MgCl$_2$ to a concentration of 1 mg/mL and placed in the fluorometer. The solution is stirred continuously using a magnetic stirrer (*see* **Note 9**).

2. Record fluorescence due to Alexa 488 (excitation 470 nm, emission 520 nm) of the aptamer probe solution at 25 °C.

3. Add ADP to a final concentration of 1–6 mM (as indicated in Fig. 3). When fluorescence is stationary add an equimolar concentration of phosphoenol pyruvate (PEP) and wait until the fluorescence is again stable.

4. Add 5 U of rabbit pyruvate kinase (PK) to the solution and wait until the fluorescence has increased to a constant level (*see* Fig. 3a, b). The increase in fluorescence is due to the formation

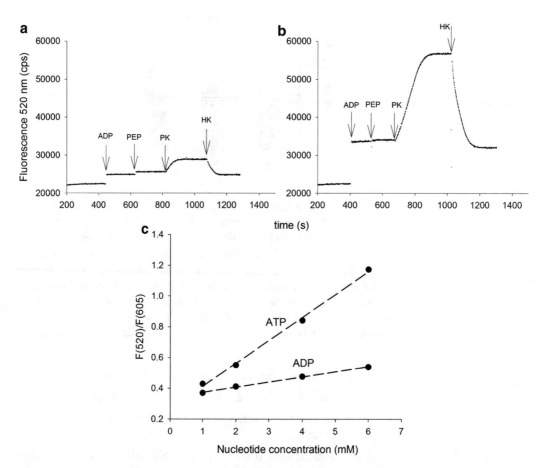

Fig. 3 Construction of calibration curves for ADP and ATP. (**a** and **b**) Fluorescence changes due to Alexa Fluor 488 on the aptamer probe (excitation 470 nm, emission 520 nm) following addition of 1 mM (**a**) or 4 mM (**b**) ADP, 1 mM (**a**) or 4 mM (**b**) phosphoenol pyruvate (PEP), 5 U pyruvate kinase (PK), and finally 5 U hexokinase (HK) to the solution containing 1 mg/mL of ATP sensor. (**c**) Resulting binding curves for nucleotide titrations calculated from the fluorescence response at 520 nm divided by the fluorescence signal recorded from the reference fluorophore (Texas Red, excitation 580 nm, emission 610 nm)

of ATP following transfer of phosphate from PEP to ADP catalyzed by pyruvate kinase. The reaction is irreversible so when it is complete all ADP has been converted to ATP.

5. Add 5 U of hexokinase and wait until the fluorescence has decreased to a stable level. The decrease in fluorescence is due to the transfer of phosphate from ATP to glucose in the cytosol buffer catalyzed by hexokinase. Also this reaction is irreversible.

6. The fluorescence following addition of ADP is taken as the ADP signal while the maximum fluorescence immediately before addition of hexokinase is taken as the ATP signal.

7. The signal of the reference fluorophore Texas Red (excitation 580 nm, emission 610 nm) is then recorded. Wait until the fluorescence is stable.

8. The calibration curves for ADP and ATP are constructed as the ADP signal divided by the reference signal and the ATP signal divided by the reference signal, respectively. These curves are shown in Fig. 3c (*see* **Note 10**).

4   Notes

1. All solutions are prepared with de-ionized water purified with a resistivity of 18.2 MΩ and sterilized by filtration with a filter pore size of 0.22 μm.

2. Ethidium bromide intercalates in DNA and is highly toxic.

3. γ-[^{32}P]-ATP is radioactive and should be handled with great care and only with appropriate protection measures.

4. Bis-/Acrylamide is a neurotoxin. Be careful and avoid direct contact.

5. The aptamer switch probe is composed of the ATP aptamer sequence and a 7-nucleotide extension, which is complementary to the 5′-end of the ATP aptamer. The construct forms a hairpin structure. By addition of a fluorophore (Alexa Flour 488) to the 5′-end of the ATP aptamer and a quencher (Black Hole I) to the 3′-end of the 7-nucleotide extension the aptamer/extension pair is converted to a signaling molecule. A polyethylene glycol extender is inserted between the ATP aptamer 3-end and the 5′-end of the 7-nucleotide extension in order to make signaling reversible.

6. Check your product on an agarose gel and add more PCR cycles if the product yield is low, then check again. Start with eight PCR cycles and, if the band on the agarose gel does not appear, do four additional cycles. If you feel that your product is not amplified any more you may add fresh DNA-polymerase.

7. How much γ-[^{32}P]-ATP should be added in depends on how radioactive the γ-[^{32}P]-ATP is.

8. Before addition of the any solution to the micro-emulsion the tip of the pipette is filled with argon 5 times and the argon is blown over the solution.

9. Calibration curves can also be constructed by adding increasing amounts of an ATP or an ADP stock solution to the nanosensor solution. However, the stock solutions of ATP and ADP must have a neutral pH. Since three of the four phosphate hydroxyl groups of ATP have pK_a values around 4.5 stock solutions prepared from adenosine 5′-triphosphate disodium hydrate become very acidic and require large amounts of NaOH to bring to neutral pH. Therefore, it is advised to prepare ATP stock solution from adenosine 5′-triphosphate di(tris) salt hydrate (tris = *tris*(hydroxymethyl)aminomethane), which makes stock solutions that are close to neutrality and therefore require only small amounts of NaOH to bring to neutral pH. Stock solutions of ADP are always close to neutrality.

10. It may be necessary to treat the nanosensors with DNase or RNase (1 U/mL) before insertion into cells. In this case it is advised to repeat the calibration after treatment with DNase or RNase.

Acknowledgements

This research was supported by EU FP7 Marie Curie program to the Initial Training Network ISOLATE. Tine Daa Schrøder was supported by the Lundbeck Foundation grant to the Nanomedicine Research Center for Cancer Stem Cell-Targeting Therapeutics (NanoCAN).

References

1. Goldbeter A (1996) Biochemical oscillations and cellular rhythms: the molecular basis of periodic and chaotic behaviour. Cambridge University Press, Cambridge

2. Noack S, Wiechert W (2014) Quantitative metabolomics: a phantom? Trends Biotechnol 32:238–244

3. Lalonde S, Eherhardt DW, Frommer WB (2005) Shining light on signaling and metabolic networks by genetically encoded biosensors. Curr Opin Plant Biol 8:574–581

4. Gu H, Lalonde S, Okumoto S, Looger LL, Scharff-Poulsen AM, Grossman AR, Kossman J, Jakobsen I, Frommer WB (2006) A novel analytical method for in vivo phosphate tracking. FEBS Lett 580:5885–5893

5. Imamura H, Nhat KP, Togawa H, Saito K, Iino R, Kato-Yamada Y, Nagai T, Noji H (2009) Visualization of ATP levels inside single living cells with fluorescence resonance energy transfer-based genetically encoded indicators. Proc Natl Acad Sci U S A 106: 15651–15656

6. Aylott JW (2003) Optical nanosensors - an enabling technology for intracellular measurements. Analyst 128:309–312

7. Lee YEK, Kopelman R (2012) Optical nanoparticle sensors for quantitative intracellular imaging. Wiley Interdiscip Rev Nanomed Nanobiotechnol 1:98–110

8. Tuerk C, Gold L (1990) Systematic evolution of ligands by exponential enrichment: RNA ligands to bacteriophage t4 DNA polymerase. Science 249:505–510

9. Ellington AD, Szostak JW (1990) In vitro selection of RNA molecules that bind specific ligands. Nature 346:818–822

10. Tang ZW, Mallikaratchy P, Yang RH, Kim YM, Zhu Z, Wang H, Tan WH (2008) Aptamer switch probe based on intramolecular displacement. J Am Chem Soc 130:11268–11269

11. Nielsen LJ, Olsen LF, Özalp VC (2010) Aptamers embedded in polyacrylamide nanoparticles: a tool for in vivo metabolite sensing. ACS Nano 4:4361–4370

12. Özalp VC, Pedersen TR, Nielsen LJ, Olsen LF (2010) Time-resolved measurements of intracellular ATP in the yeast *Saccharomyces cerevisiae* using a newtype of nanobiosensor. J Biol Chem 285:37579–37588

13. Ytting CK, Fuglsang AT, Hiltunen JK, Kastaniotis AJ, Özalp VC, Nielsen LJ, Olsen LF (2012) Measurements of intracellular ATP provide new insight into the regulation of glycolysis in the yeast *Saccharomyces cerevisiae*. Integr Biol 4:99–107

14. Jing M, Yingmiao L, Zahid NR, Zhongguang Y, Johannes HU, Bruce AS, Bryan MC (2010) In vivo selection of tumor-targeting RNA motifs. Nat Chem Biol 6:22–24

Chapter 2

SELEX of Cell-Specific RNA Aptamers

Katharina Berg, Eileen Magbanua, and Ulrich Hahn

Abstract

This chapter focuses on the selection of RNA aptamers, which bind to specific cell surface components and thus can be internalized receptor mediated. Such aptamers discriminate between different tissues, e.g., detect malignant cells, and target them or induce apoptosis through drug internalization. However, before starting the selection process the choice of an ideal target can be challenging. To give an example for the selection of cell specific aptamers, we here used the interleukin-6 receptor (IL-6R) as a target, which is presented on hepatocytes, neutrophils, monocytes, and macrophages.

Key words Interleukin-6 receptor, SELEX, Internalization, Delivery

1 Introduction

Cell surfaces exhibit a high amount of diverse lipids, proteins, and carbohydrates, whereas their composition differs tremendously between various cell types. Therefore aptamers can be selected for a variety of targets like growth factors, adhesion molecules, and cell surface receptors which are exclusively presented on special cell types [1]. Such aptamers can differentiate between cell types, for example healthy and tumor cells or even different types of cancer cells [2–5]. Therefore cell-specific aptamers have a high potential for diagnosis and as therapeutics.

By targeting a cell surface molecule that underlies natural internalization processes, an aptamer can act as a molecular carrier to convey fluorophores, chemotherapeutics, or other agents into cells [3, 6–9]. A prerequisite is that binding of the aptamer (including its cargo) does not interfere with the natural transport mechanism of the receptor.

In some cases, aptamer binding can interfere with receptor-ligand interaction or the receptor recycling processes, thus inhibiting the binding of natural ligands and/or altering the subsequent signal transduction [10]. However, recycling processes of most cell surface lipids, receptors, or tumor markers are still not completely

Günter Mayer (ed.), *Nucleic Acid Aptamers: Selection, Characterization, and Application*, Methods in Molecular Biology, vol. 1380, DOI 10.1007/978-1-4939-3197-2_2, © Springer Science+Business Media New York 2016

understood. Though there are some reports on internalized cell specific aptamers, the localization of both aptamer and delivered cargo cannot be predicted and determined easily and differs depending on the respective receptor and cargo.

One example of such an aptamer represents PSM-A10, an RNA aptamer, which inhibits the prostate-specific membrane antigen (PSMA) with a K_i of 11.9 nM. PSM-A10 is able to bind LNCaP human prostate cancer cells presenting PSMA, but not PSMA-devoid PC-3 human prostate cancer cells [4]. It can be used for the localized delivery of siRNA molecules, polymer nanoparticles, or toxins like gelonin [6–8].

Another example is an RNA aptamer named AIR-3 as well as its truncated form AIR-3A, which interact with high affinity (K_d = 20 nM) and specificity with the human interleukin-6 receptor (IL-6R) [11]. AIR-3A is also able to deliver therapeutic agents, such as chlorin e6 (ce6), into its target cell. This enables a specific ce6-mediated photodynamic therapy [12]. AIR-3 even might be used for targeted chemotherapy when intrinsically comprising the cytostaticum 5-fluoro-2′-deoxy-uridine [13].

The selection of aptamers consists of three main steps: The first step is the incubation of an RNA library with the target molecule. This library consists of around 10^{14} different sequences, which share a randomized region of 30–50 nucleotides (N30–50). This region is flanked by two constant regions for T7-transcription and RT-PCR.

After incubation, bound and unbound oligonucleotides are separated. In a common in vitro selection procedure, the target molecule is immobilized on a column or beads, while other protocols utilize gel electrophoresis or nitrocellulose membrane filtration [14, 15]. In the last step of each SELEX cycle, bound molecules are eluted, amplified via RT-PCR, and then transcribed to generate an enriched pool of RNA oligonucleotide species for the next selection round. This iterative process is repeated 5–15 times, followed by cloning and sequencing of the dsDNA library [16, 17].

Obtained oligonucleotides can be characterized for their binding abilities and their specificity either by protein-RNA interaction assays like filter retention, electric mobility shift, and plasmon resonance assays or directly on cells using flow cytometry or confocal microscopy. Subsequently to the SELEX procedure, the affinity of the aptamers to cells has to be determined. Therefore, the aptamer is fluorescently labelled. Adequate control cells omitting the target protein are required to confirm specificity of the aptamers and allow characterizations (e.g., toxicity, inhibiting or activating events, aptamer internalization).

The interleukin-6 receptor, used as an aptamer target by our group, interacts with its cytokine interleukin-6 (IL-6) and two glycoproteins 130 (gp130). The resulting gp130 dimerization initiates an intracellular signal transduction, in which event the expression of various genes is activated [18, 19]. We used Ba/F3/gp130/IL6R/TNF cells to analyze the influence of the aptamer on signal transduction and examined receptor mediated endocytosis. As a control cell line Ba/F3/gp130/TNF cells without IL-6R were used.

Generally, before starting a SELEX there are some aspects, which have to be taken into account. In most cases, it is indispensable to use a soluble form of the target, often corresponding to the extracellular domain of a cell surface protein if that has been chosen. Nevertheless it needs to be ensured, that the protein is correctly folded and fully functional. If the cell surface protein of choice cannot be produced in its soluble form, cell-SELEX [20–22] would be a favorable alternative. For a successful SELEX experiment, it is desirable to have as much information as possible about the structure as well as the physical and chemical properties of the target protein. This is required to assess ideal buffer conditions, immobilization of the target and an appropriate pre- or counterselection molecule to enhance the selectivity. Some chosen targets might be highly glycosylated or some just might be difficult to purify, rendering selection more challenging.

If the selected RNA aptamer is planned to be used in vivo, it might be necessary or at least advantageous to enhance its stability. Post-selective modifications can interfere with aptamer folding and target interaction, it is therefore recommended to use a stabilized RNA library (e.g., with 2′-O-methyl or 2′-fluoro modifications) directly in the SELEX process [21, 23–25]. For certain applications, DNA aptamers might be preferential [26].

2   Materials

Prepare all solutions RNase-free (*see* **Note 1**).

2.1 Preparation of RNA Library and In Vitro T7 Transcription

1. 100 µM single strand DNA library: 5′-AATGCTAATACGACT CACTATAGG AAGAAAGAGGTCTGAGACATTCT–N50– CTTCTGGAGTTGACGTTGCTT-3′ (*see* **Note 2**).

2. 100 µM RT primer: 5′-AAGCAACGTCAACTCCAGAAG-3′

3. Klenow Fragment (5 U/µL), 10× Klenow Fragment buffer: 500 mM Tris-HCl (pH 8.0 at 25 °C), 50 mM MgCl$_2$, 10 mM DTT.

4. 10× hybridization buffer: 200 mM Tris-HCl, pH 8.0, 500 mM NaCl, 10 mM EDTA.

5. dNTP mix: 25 mM ATP, 25 mM CTP, 25 mM GTP, 25 mM TTP.

6. 20 U/µL T7 RNA Polymerase (*see* **Note 3**).

7. 5× transcription buffer: 200 mM Tris-HCl, pH 7.9.

8. 100 mM MgCl$_2$.

9. NTP-mix A: 25 mM ATP, 25 mM CTP, 25 mM GTP, 25 mM UTP (*see* **Note 4**).

2.2 Biotinylation of Target Protein

1. 100 μg soluble IL-6R (sIL_6R) (Conaris).

2. 2 mg *Sulfo*-NHS-LC-Biotin.

3. Phosphate-buffered saline (PBS): 0.14 M NaCl, 2.7 mM KCl, 1.5 mM KH$_2$PO$_4$, 6.5 mM NaH$_2$PO$_4$, pH 7.4.

4. SELEX buffer: 3 mM MgCl$_2$ in PBS (*see* **Note 5**).

5. Slide-A-Lyzer® dialysis cassette (MWCO 10K; Thermo Fisher Scientific) (*see* **Note 6**).

6. Nitrocellulose membrane.

7. ExtrAvidin® Alkaline phosphatase.

8. Blocking buffer: 5 % (w/v) skim milk powder in PBS.

9. AB buffer: 2.5 % (w/v) skim milk powder in PBS.

10. Reaction buffer: 100 mM Tris-HCl, pH 9.5, 4 mM MgCl$_2$.

11. AP solution: 0.01 % (w/v) 5-bromo-4-chloro-3-indoxyl phosphate, 0.002 % (w/v) nitro blue tetrazolium chloride in reaction buffer.

2.3 Immobilization on Streptavidin-Coated Dynabeads

1. Dynabeads® M-280 Streptavidin (Invitrogen).

2. Magna-Sep™ Magnetic Particle Separator.

3. SELEX buffer: 3 mM MgCl$_2$ in PBS.

4. BSA solution: 20 μg/μL in water.

5. Coupling buffer A: 1 μg BSA/μL, 1× PBS, pH 7.4.

6. Coupling buffer B: 1 μg BSA/μL, 1.25× PBS, pH 7.4.

2.4 Reverse Transcription-Polymerase Chain Reaction

1. 5 U/μL FIREPol® DNA Polymerase (Solis BioDyne).

2. 10× PCR reaction buffer B: 0.8 M Tris-HCl, 0.2 M (NH$_4$)$_2$SO$_4$, 0.2 % w/v Tween-20 (Solis BioDyne).

3. 15 U/μL SuperScript™ III Reverse Transkriptase.

4. 5× cDNA synthesis buffer: 250 mM Tris-acetate, pH 8.4, 375 mM potassium acetate, 40 mM magnesium acetate.

5. 25 mM MgCl$_2$.

6. 100 mM DTT.

7. dNTPmix: 25 mM ATP, 25 mM CTP, 25 mM GTP, 25 mM TTP.

8. 100 μM RT primer: 5′-AAGCAACGTCAACTCCAGAAG-3′.

9. 100 μM T7 primer: 5′-<u>AATGCTAATACGACTCACTATAG</u>GAAGAAAGAGGTC TGAGACATT-3′.

2.5 Cloning	1. TOPO TA Cloning Kit (pCR2.1).
2.6 Filter Retention Assay	1. [α-^{32}P]-ATP (3000 Ci/mmol, 10 mCi/mL, Hartmann Analytic).
	2. NTP-mix B: 25 mM ATP, 25 mM CTP, 25 mM GTP, 25 mM UTP.
	3. SELEX buffer: 3 mM MgCl$_2$ in PBS.
	4. Aminohexanoic acid buffer: 20 % (v/v) Methanol; 40 mM 6-aminohexanoic acid.
	5. Filtering paper (e.g., Rotilabo®-Blottingpaper, Carl Roth).
	6. 500 µM sIL-6R (Conaris).
	7. Nitrocellulose membrane (0.45 µm).
	8. Minifold® I dot-blot-system (e.g., Schleicher & Schuell).
2.7 Flow Cytometry	1. 10 µM fluorescently labeled RNA (*see* **Note 7**).
	2. Mouse anti-human Interleukin-6 receptor antibody (Bender MedSystems) (*see* **Note 7**).
	3. Mouse anti-human gp130 antibody (R&D Systems) (*see* **Note 8**).
	4. Goat anti-mouse Ig–APC (BD Pharmingen).
	5. Ba/F3/gp130/IL6R/TNF cells.
	6. Ba/F3/gp130/TNF cells.
	7. SELEX buffer: 3 mM MgCl$_2$ in PBS.
	8. FACSCalibur flow cytometer (BD Bioscience).

3 Methods

3.1 Preparation of RNA Library	1. Heat 2.5 µL ssDNA library (100 µM) with 5 µL RT primer (100 µM), 5 µL hybridization buffer (10×), and 37.5 µL water to 95 °C for 5 min and cool down slowly to room temperature afterwards.
	2. Mix 50 µL hybridization product with 1 µL Klenow Fragment (10 U/µL), 10 µL Klenow Fragment buffer, 2 µL dNTPs (25 mM), and 37 µL water and incubate for 1 h at 37 °C. Then inactivate Klenow Fragment by heating to 75 °C for 10 min.
	3. For in vitro transcription, mix 20 µL double-strand DNA library with 20 µL transcription buffer (5×), 15 µL MgCl$_2$ (100 mM), 10 µL NTP-Mix (25 mM), 1 µL T7 RNA Polymerase (25 U/µL), and 34 µL water and incubate for 30 min at 37 °C.
3.2 Biotinylation of Target Protein	1. Dissolve 2 mg Sulfo-NHS-LC-Biotin in 360 µL water (yielding a 10 mM solution, which needs to be diluted to 1 mM) (*see* **Note 9**).

2. Mix 100 μL sIL-6R (1.15 μg/μL) with 8.8 μL freshly prepared Sulfo-NHS-LC-Biotin reagent (1 mM) and incubate on ice for 15 min. Afterwards incubate for additional 15 min at room temperature (*see* **Note 10** and **11**).

3. Remove excess of non-reacted and hydrolyzed biotin reagent by dialysis with Slide-A-Lyzer® dialysis cassette. Therefore, fill mixture into the dialysis cassette and dialyze in 500 mL stirring SELEX buffer for 1 h at 4 °C. Then exchange the buffer and dialyze the mixture overnight.

4. Verify biotinylation by dot blot analysis. Therefore, place 3 μL of biotinylated protein on a nitrocellulose membrane and 3 μL non-biotinylated protein next to it. Let drops dry up.

5. Place the membrane in 25 mL blocking buffer for 1 h at room temperature and swing gently. Next, wash membrane briefly in 25 mL PBS, place the membrane in 25 mL AB buffer with 10 μL ExtrAvidin® Alkaline phosphatase for 1 h at room temperature, and swing gently again.

6. Wash the membrane twice in 25 mL PBS and once in 25 mL reaction buffer for 5 min at room temperature, respectively.

7. Add 25 mL reaction buffer to the membrane including 12.5 μL AP solution and incubate until color reaction significantly appears. To stop reaction transfer the membrane into water.

3.3 Immobilization on Streptavidin-Coated Dynabeads

1. Transfer 500 μL Dynabeads® into a reaction tube and put it into the magnetic particle separator for magnetic separation (*see* **Note 10**). Discard supernatant, take reaction tube out of separator, and resolve the beads in 1 mL coupling buffer A (*see* **Note 12**). Repeat this washing step four times.

2. Add 100 μg biotinylated protein to the beads and incubate for 15 min at 4 °C. Wash the beads five times with 500 μL coupling buffer A and then twice with 500 μL coupling buffer B.

3. Store the beads in 1.5 mL coupling buffer B at 4 °C.

3.4 SELEX

1. For the initial round of selection, mix 500 pmol of the RNA library with 100 pmol target protein immobilized on magnetic beads in SELEX buffer and incubate for 30 min at 37 °C. Discard unbound RNA by magnetic separation (*see* Subheading 3.3) and wash beads with 200 μL SELEX buffer (*see* **Note 13**).

2. Elute bound RNA by resuspending magnetic beads in 50 μL water and heat for 3 min at 80 °C. Then separate RNA containing water from magnetic beads by magnetic separation and transfer supernatant to a new reaction tube for reverse transcription-polymerase chain reaction (RT-PCR).

3. For RT-PCR mix the following components: 50 μL eluate from **step 2**, 10 μL 10× PCR reaction buffer B, 4 μL 5× cDNA

synthesis buffer, 6 µL 25 mM $MgCl_2$, 1.2 µL dNTPmix, 1 µL RT primer, 1 µL T7 primer, 2 µL DTT, and 22.8 µL water. Incubate samples at 65 °C for 5 min and cool down to room temperature afterwards.

4. Add 1 µL SuperScript™ III Reverse Transcriptase and 1 µL FIREPol® DNA polymerase and incubate reaction mix for 10 min at 54 °C for reverse transcription.

5. Use reaction mixture from **step 4** instantly for PCR amplification comprising following steps: denaturing for 30 s at 95 °C, annealing for 30 s at 60 °C and elongation for 30 s at 72 °C and appropriate number of PCR cycles (*see* **Note 14**). Verify PCR product by performing standard polyacrylamide gel electrophoresis.

6. Transcribe resulting double stranded DNA (dsDNA) of **step 5** into RNA. Therefore, mix 10 µL dsDNA with 20 µL 5× transcription buffer, 15 µL $MgCl_2$, 10 µL NTP-mix A, 1 µL T7 RNA polymerase and 44 µL water and incubate for 30 min at 37 °C.

7. Verify RNA by performing polyacrylamide gel electrophoresis.

8. Start new selection round with 20 µL obtained RNA (*see* **Note 15**).

9. To investigate individual clones, clone dsDNA library via TOPO TA cloning (*see* **Note 16**) and perform sequence analysis.

3.5 Filter Retention Assay

1. Perform T7 transcription to radioactively label RNA. Therefore, mix 10 µL dsDNA with 20 µL 5× transcription buffer, 15 µL $MgCl_2$, 10 µL NTP mix B, 1 µL [α-^{32}P]-ATP, and 1 µL T7 RNA polymerase with 43 µL water and incubate overnight at 37 °C.

2. Isolate RNA product by gel purification and dissolve RNA in 20 µL of water.

3. Mix 1 µL of radioactively labeled RNA (<1 nM) with 199 µL SELEX buffer.

4. Dilute target protein in SELEX buffer to a concentration of 5 µM in a total volume of 12 µL (*see* **Note 17**). For serial protein dilution mix 6 µL 5 µM protein with 6 µL SELEX buffer and continue to dilute the protein four times.

5. Start binding assay: mix 5 µL of each protein dilution with 20 µL RNA (**step 3**) and incubate for 30 min at 22 °C. As negative control, dilute 20 µL RNA with 5 µL SELEX buffer (*see* **Note 18**).

6. Cut one filter paper and nitrocellulose membrane to the size of the dot blot apparatus. Equilibrate nitrocellulose membrane in aminohexanoic acid buffer for 30 min at room temperature. Afterwards, equilibrate both nitrocellulose membrane and filter paper in SELEX buffer for at least 5 min.

7. Assemble dot blot apparatus. Therefore, place filter paper beneath the nitrocellulose membrane inside dot blot apparatus.

8. Apply vacuum on dot blot apparatus and wash each well twice with 200 µL SELEX buffer.

9. Transfer all samples into the dot blot apparatus in respective wells and let the samples completely pass through the membrane (*see* **Note 19**).

10. Wash each well twice with 200 µL SELEX buffer and let buffer completely pass through the membrane. Then take out the nitrocellulose membrane and let it dry.

11. Prepare 100 % controls. Mix 5.4 µL SELEX buffer with 0.6 µL radioactively labeled RNA. Apply twice 2 µL to the membrane and let it dry.

12. Transfer membrane into a plastic bag and expose properly until analysis.

13. For analysis use a phosphor imager.

14. Quantify intensities of the dots and calculate dissociation constant (Fig. 1).

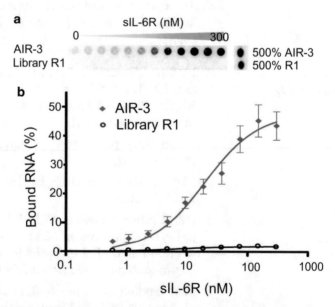

Fig. 1 (**a**) Filter retention assay. Constant amounts (<1 nM) of ^{32}P-radioactively labeled aptamer AIR-3 or RNA starting library R1 were incubated with increasing amounts of sIL-6R (0–300 nM). Protein-bound RNA could be visualized by autoradiography. To quantify the amount of RNA a 500 or 100 % control is used. (**b**) To calculate dissociation constants, fractions of bound RNA molecules (AIR-3A, *red*; library, *black*) were plotted against the concentrations of sIL-6R (logarithmic scale). Data points represent mean values of ten independent measurements (further details *see* [11])

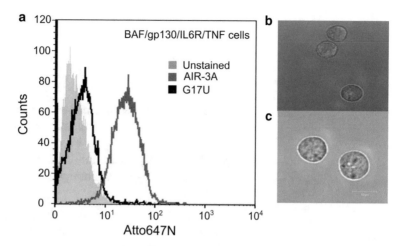

Fig. 2 (**a**) Flow cytometry analysis. Upon binding of Atto-647 N-labeled Aptamer AIR-3A (*red*) to BAF/gp130/IL6R/TNF cells presenting the human IL-6 receptor, fluorescence shifts to higher values. Cells incubated with the variant G17U (Atto647N-labeled control, *black*) resulted in a smaller shift compared to AIR-3A. Laser scanning microscopic image of AIR-3A (**b**) or its variant G17U (**c**) with BAF/gp130/IL6R/TNF cells. The aptamer (**b**) is internalized into these cells (further details *see* [11])

3.6 Flow Cytometry

Perform all steps at 4 °C and use SELEX buffer pre-cooled to 4 °C (*see* **Note 20**).

1. Count cells and use 1×10^5 cells for each sample.

2. Wash cells twice. Therefore, pellet cells for 3 min at $800 \times g$ and resuspend with 350 μL SELEX buffer (*see* **Note 21** and **22**).

   (a) Control cells are not further treated

   (b) Aptamer binding: Add 1 μL fluorescently labeled RNA aptamer (10 pmol) to the cells and incubate for 30 min on ice.

   (c) IL-6R and gp130 verification: Add either 1 μL IL-6R antibody or gp130 antibody to the cells and incubate for 30 min on ice. Then wash cells once with SELEX buffer and add 1 μL anti-mouse Ig–APC antibody to the cells and incubate for another 30 min on ice.

3. Wash cells once with 350 μL SELEX buffer.

4. Determine fluorescence intensities by flow cytometry analysis and compare intensities (Fig. 2).

4 Notes

1. Aqueous solutions can be treated with diethylpyrocarbonate (DEPC) to inactivate RNases due to acetylation of histidine, lysine, cysteine and tyrosine residues. Therefore solution is

treated with 0.1 % (v/v) DEPC and stirred overnight. Autoclaving afterwards leads to destruction of DEPC. Tris buffered solutions should not be treated with DEPC as it inactivates DEPC. Alternatively, RNase-free water derived from special filtration plants can be used.

2. Given are sequences of a library used for the selection of aptamers binding to the human interleukin-6 receptor [11]. Randomized region can be optimized regarding length and given secondary structure (e.g., inserting a stem loop region). Optimal lengths of randomized region are between 20 and 50 nucleotides.

3. Instead of wild type RNA polymerase, variants (e.g., RNAP Y639F or R425C) can be used which are able to incorporate modified nucleotides.

4. Unmodified standard nucleotides can be replaced by modified nucleotides (e.g., 2'-fluoronucleotides, *O*-methyl nucleotides), which require the usage of specialized RNA polymerases. Mostly the reaction time needs to be increased to 4 h instead of 30 min.

5. Alternatively, other dialysis cassettes, membranes, tubes, and suchlike can be used.

6. If the target protein exhibits a very low isoelectric point the use of positive ions, e.g., calcium ions are recommended.

7. Fluorescently labeled RNA can be purchased or labeled by inserting fluorescing nucleotide analogs during T7 transcription (e.g., guanosine-5'-*O*-(3-thiotriphosphate)).

8. Choose antibodies which are applicable for FACS measurements.

9. There are several types of magnetic beads to choose from. In the given example streptavidin coated beads were used, but also the chemical coupling to carboxyl beads is possible.

10. Prepare *Sulfo*-NHS-LC-Biotin reagent immediately before usage.

11. *Sulfo*-NHS-LC-Biotin reagent should be used at threefold molar excess to the protein.

12. Dynabeads® are paramagnetic polystyrene particles that can be separated by a magnet. Placing a reaction tube which contains Dynabeads® into the Magnetic Particle Separator always leads directly to magnetic separation. Successful separation can be observed when solution cleared up.

13. Prevent the magnetic beads to become dry.

14. In the course of selection, the number of washing steps is increased to favor the selection of high affinity nucleic acids. So start with one washing step and add one step in each selection round.

15. The optimal number of PCR cycles lies between 4 and 20. To prevent formation of PCR side products emulsion PCR (e.g.,

Micellula DNA Emulsion & Purification Kit by EUR$_X$) can be performed alternatively [27].

16. Stop SELEX procedure after 10–15 rounds. The number of selection rounds depends on the affinity of the nucleic acid pool to the target. If the whole library is analyzed by next generation sequencing less round are required.

17. To increase the enrichment of nucleic acids which are only binding to the target molecule and not to Dynabeads® or Streptavidin, it is common to do a preSELEX prior to each SELEX round. Therefore use Dynabeads® with only biotin and incubate them at 37 °C for 30 min with the prior round first. Then use the supernatant as described above in the actual SELEX round. In this step it is also possible to use Dynabeads® coupled with non-target molecules to exclude those nucleic acids which are not binding specifically.

18. Using a Taq polymerase for polymerase chain reaction leads to adenine overhangs at the 3′-end of the PCR product. These overhangs are required for successful TA cloning. Alternatively to the mentioned TOPO TA cloning kit, a vector can be used which exhibits 5′-thymine overhangs after restriction (e.g., endoribonuclease XcmI), or standard cloning can be performed at restriction sites which were added during polymerase chain reaction or presented within the primer sites.

19. 12 μL 5 μM protein solution is needed to perform one binding assay with one sample of RNA. If several aptamers are analyzed simultaneously, protein amount has to be increased proportionally.

20. End concentrations of protein are 1000, 0.500, 0.250, 0.125, 0.063, and 0.031 μM.

21. Avoid touching the nitrocellulose membrane with the tip. The membrane can be damaged and distort results.

22. Resuspend cells gently, prevent formation of foam, and carry out all steps quickly without delay to prevent cell death.

References

1. Mager MD, LaPointe V, Stevens MM (2011) Exploring and exploiting chemistry at the cell surface. Nat Chem 3:582–589
2. Davis KA, Lin Y, Abrams B, Jayasena SD (1998) Staining of cell surface human CD4 with 2′-F-pyrimidine-containing RNA aptamers for flow cytometry. Nucleic Acids Res 26:3915–3924
3. Li N, Ebright JN, Stovall GM, Chen X, Nguyen HH, Singh A, Syrett A, Ellington AD (2009) Technical and biological issues relevant to cell typing with aptamers. J Proteome Res 8:2438–2448
4. Lupold SE, Hicke BJ, Lin Y, Coffey DS (2002) Identification and characterization of nuclease-stabilized RNA molecules that bind human prostate cancer cells via the prostate-specific membrane antigen. Cancer Res 62:4029–4033
5. Shangguan D, Tang Z, Mallikaratchy P, Xiao Z, Tan W (2007) Optimization and modifications of aptamers selected from live cancer cell lines. Chembiochem 8:603–606

6. Chu TC, Marks JW 3rd, Lavery LA, Faulkner S, Rosenblum MG, Ellington AD, Levy M (2006) Aptamer:toxin conjugates that specifically target prostate tumor cells. Cancer Res 66:5989–5992

7. Chu TC, Twu KY, Ellington AD, Levy M (2006) Aptamer mediated siRNA delivery. Nucleic Acids Res 34:e73

8. Farokhzad OC, Jon S, Khademhosseini A, Tran TN, Lavan DA, Langer R (2004) Nanoparticle-aptamer bioconjugates: a new approach for targeting prostate cancer cells. Cancer Res 64:7668–7672

9. Ray P, Cheek MA, Sharaf ML, Li N, Ellington AD, Sullenger BA, Shaw BR, White RR (2012) Aptamer-mediated delivery of chemotherapy to pancreatic cancer cells. Nucleic Acid Ther 22:295–305

10. Kraus E, James W, Barclay AN (1998) Cutting edge: novel RNA ligands able to bind CD4 antigen and inhibit CD4+ T lymphocyte function. J Immunol 160:5209–5212

11. Meyer C, Eydeler K, Magbanua E, Zivkovic T, Piganeau N, Lorenzen I, Grotzinger J, Mayer G, Rose-John S, Hahn U (2012) Interleukin-6 receptor specific RNA aptamers for cargo delivery into target cells. RNA Biol 9:67–80

12. Kruspe S, Meyer C, Hahn U (2014) Chlorin e6 conjugated interleukin-6 receptor aptamers selectively kill target cells upon irradiation. Mol Ther Nucleic Acids 3:e143

13. Kruspe S, Hahn U (2014) An aptamer intrinsically comprising 5-fluoro-2′-deoxyuridine for targeted chemotherapy. Angew Chem Int Ed Engl 53:10541–10544

14. Gopinath SC (2007) Methods developed for SELEX. Anal Bioanal Chem 387:171–182

15. Jing M, Bowser MT (2013) Tracking the emergence of high affinity aptamers for rhVEGF165 during capillary electrophoresis-systematic evolution of ligands by exponential enrichment using high throughput sequencing. Anal Chem 85:10761–10770

16. Gu G, Wang T, Yang Y, Xu X, Wang J (2013) An improved SELEX-Seq strategy for characterizing DNA-binding specificity of transcription factor: NF-kappaB as an example. PLoS One 8:e76109

17. Schutze T, Wilhelm B, Greiner N, Braun H, Peter F, Morl M, Erdmann VA, Lehrach H, Konthur Z, Menger M, Arndt PF, Glokler J (2011) Probing the SELEX process with next-generation sequencing. PLoS One 6:e29604

18. Guo Y, Xu F, Lu T, Duan Z, Zhang Z (2012) Interleukin-6 signaling pathway in targeted therapy for cancer. Cancer Treat Rev 38:904–910

19. Ozbek S, Grotzinger J, Krebs B, Fischer M, Wollmer A, Jostock T, Mullberg J, Rose-John S (1998) The membrane proximal cytokine receptor domain of the human interleukin-6 receptor is sufficient for ligand binding but not for gp130 association. J Biol Chem 273:21374–21379

20. Mayer G, Ahmed MS, Dolf A, Endl E, Knolle PA, Famulok M (2010) Fluorescence-activated cell sorting for aptamer SELEX with cell mixtures. Nat Protoc 5:1993–2004

21. Meyer C, Hahn U, Rentmeister A (2011) Cell-specific aptamers as emerging therapeutics. J Nucleic Acids 2011:904750

22. Sefah K, Shangguan D, Xiong X, O'Donoghue MB, Tan W (2010) Development of DNA aptamers using Cell-SELEX. Nat Protoc 5:1169–1185

23. Ibach J, Dietrich L, Koopmans KR, Nobel N, Skoupi M, Brakmann S (2013) Identification of a T7 RNA polymerase variant that permits the enzymatic synthesis of fully 2′-O-methyl-modified RNA. J Biotechnol 167:287–295

24. Padilla R, Sousa R (2002) A Y639F/H784A T7 RNA polymerase double mutant displays superior properties for synthesizing RNAs with non-canonical NTPs. Nucleic Acids Res 30:e138

25. Meyer C, Berg K, Eydeler-Haeder K, Lorenzen I, Grotzinger J, Rose-John S, Hahn U (2014) Stabilized Interleukin-6 receptor binding RNA aptamers. RNA Biol 11:57–65

26. Faryammanesh R, Lange T, Magbanua E, Haas S, Meyer C, Wicklein D, Schumacher U, Hahn U (2014) SDA, a DNA aptamer inhibiting E- and P-selectin mediated adhesion of cancer and leukemia cells, the first and pivotal step in transendothelial migration during metastasis formation. PLoS One 9:e93173

27. Shao K, Ding W, Wang F, Li H, Ma D, Wang H (2011) Emulsion PCR: a high efficient way of PCR amplification of random DNA libraries in aptamer selection. PLoS One 6:e24910

Chapter 3

Developing Aptamers by Cell-Based SELEX

Silvia Catuogno, Carla Lucia Esposito, and Vittorio de Franciscis

Abstract

The reliable targeting of cell surface disease-associated proteins is a major challenge in chemical biology and molecular medicine. In this regard, aptamers represent a very attractive and innovative class of ligand molecules. Aptamers are generated by a reiterated in vitro procedure, named SELEX (Systematic Evolution of Ligands by Exponential enrichment). In order to generate aptamers for heavily modified cell surface-bound proteins and transmembrane receptors, the SELEX procedure has been recently adapted to the use of living cells as complex targets (referred as "cell-SELEX").

Here we give an overview on the most recent advances in the field of cell-SELEX technology, providing a detailed description of the differential cell-SELEX approach that has been developed in our laboratory to identify specific signatures for human malignant glioma and non-small-cell lung cancer. The procedures used for the evaluation of binding specificity and for the preliminary identification of potential target receptors will be also described.

Key words Cell-SELEX, Aptamer, NSCLC, Glioma, RTK

1 Introduction

1.1 Cell-Based SELEX

Several important human diseases, including cancer and metabolic diseases are characterized by the presence of various alterations in cell-surface proteins. Changes in the expression level, localization or structural changes are frequently the cause of abnormal intracellular signaling ultimately determining the pathological states. Therefore, the specific targeting of disease-associated membrane proteins has recently become a challenge for the development of new therapeutic or diagnostic tools.

In this perspective, nucleic acid aptamers are revealing highly promising tools to identify and target cell-surface proteins [1]. Aptamers are short single stranded DNA/RNA molecules that resemble antibodies in many ways. By folding into complex tertiary structures, these oligonucleotides bind with high affinity and specificity their target proteins, thus often leading to the modulation of their activity [2].

Günter Mayer (ed.), *Nucleic Acid Aptamers: Selection, Characterization, and Application*, Methods in Molecular Biology, vol. 1380, DOI 10.1007/978-1-4939-3197-2_3, © Springer Science+Business Media New York 2016

Aptamers show many advantages over proteins or antibodies:
(1) they are chemically synthesized avoiding the use of animal cells,
thus providing high batch-to-batch fidelity; (2) they can be easily
modified to enhance their stability, bioavailability and pharmacoki-
netics; (3) depending on their formulation, aptamers are usually
poorly or not immunogenic [3].

Different approaches have been adopted to generate aptamers
for cell surface molecular targets. In many cases, soluble purified
cell surface proteins have been used as targets for aptamer selection
in vitro (protein-SELEX) [4–7]. Although this approach shows the
advantage to be conducted under well-controlled conditions, the
selection is performed in a non-physiological context, leading to
the possibility that the selected aptamer might not recognize the
same target in its native conformation.

The application of the SELEX technology using whole living
cells (cell-SELEX) as complex target allows to overcome this
problem by selecting aptamers under native conditions in a
physiological context. Globally three variants of cell-SELEX have
been adopted that permit to achieve different objectives (Fig. 1).

A first strategy (Fig. 1a) allows to select aptamers against a
previously identified target protein, by using in the positive selection
step an engineered cell line forced to express the recombinant tar-

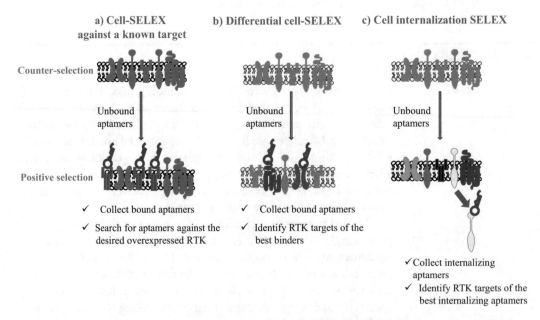

Fig. 1 Schematic representation of three variant of cell-based SELEX proposed to achieve different objectives:
(**a**) protocol to select aptamers against a known target; (**b**) protocol to identify a specific cell signature without
a prior knowledge of the target protein; (**c**) protocol to specifically select cell internalizing aptamers

get protein. To avoid the parallel enrichment of aptamers for unwanted targets, the introduction at each round of one (or more) counter-selection steps with parental cells is required. To date, several groups have successfully adopted this strategy to generate aptamers that specifically bind cell surface receptors.

An innovative cell-SELEX strategy has been successfully applied to develop aptamers that specifically bind the transforming growth factor-β type III receptor (TGFβRIII) [8], ectopically expressed on the surface of Chinese hamster ovary (CHO) cells, as well as to generate a high-affinity ssDNA aptamer that specifically binds to the HCV E2 envelope glycoprotein [9]. This technique offers the possibility to favor the selection of target-specific aptamers since the selective pressure is increased by the overexpression of the target protein coupled to the use of an appropriate counter-selection procedure.

An alternative strategy (referred to as "differential cell-SELEX") (Fig. 1b) allows targeting of a specific cell type without any prior knowledge of the target protein, thus leading to the identification of multiple ligands able to discriminate among even closely related cell phenotypes. In this case, two different cell lines, or the same cell line under different growth conditions or insults, are used in the positive and in the counter-selection steps. The enrichment for nonspecific ligands can be avoided by modulating the stringency at each SELEX round in different ways, including the use of suitable competitors during the incubation step, the reduction of target concentration and the increase in the number of washings after incubation.

Shangguan et al. applied such approach to identify ssDNA aptamers that by specifically binding T-cell acute lymphocytic leukemia (ALL) cells are able to distinguish them from B-cell lymphoma cells [10]. Moreover, in the same laboratory Chen et al. developed a panel of aptamers specific for small lung cancer (SCLC) that is the most aggressive subtype of lung cancer with a very short life expectancy. They isolated a set of oligonucleotides able to discriminate small lung cancer cells versus large cell lung cancer, providing a specific tumor signature for early diagnosis of SCLC [11]. A similar approach has been used by Sefah et al. to identify aptamers specific for colorectal cancer versus normal tissue or other cancer cells [12].

The specific protocols adopted by different laboratories may largely vary in terms of the specific scheme of SELEX applied (for example the alternation of the selection and counter-selection steps), the cell types used in the positive and counter-selection steps, the time and the temperature of incubation of the library with the cells and the stringency applied through the SELEX rounds.

In order to develop more effective delivery tools for therapeutics, a variant of the two cell-based SELEX strategies (Fig. 1a, b) has been introduced in order to enrich for aptamers capable of selective internalization (Fig. 1c). The protocol (referred as "cell internalization SELEX") includes ice-cold 0.5 M NaCl in Dulbecco's Phosphate-Buffered Saline (DPBS) washings (High Salt Wash) to remove surface-bound aptamers that do not internalize into target cells [13, 14].

1.2 Differential Cell-SELEX on Glioma and Lung Cancer Cells

In our laboratory, we adopted differential cell-SELEX to generate a panel of high affinity RNA aptamers directed against human malignant glioma [15] and non-small-cell lung cancer (NSCLC) cells surface antigens [16].

Gliomas are the most common primary malignant brain tumors characterized by variable grade of malignancy and histological features [17–22]. With the intention to identify new specific molecular markers preferentially expressed on the surface of the more malignant glioma phenotype, we used human U87MG glioma cell line as target for the selection step and the less aggressive human T98G glioma cells for the counter-selection steps [15]. These two glioma cell lines differ for the potential to form tumors in nude mice, being highly tumorigenic and poorly tumorigenic respectively. We used a library of 2'-fluoro pyrimidines (2'-F)-RNAs and performed 14 rounds of SELEX, progressively increasing the selective pressure by changing both incubation and washing conditions. Finally, we identified ten families of highly related aptamers that cover more than 46 % of all individual sequences obtained. Among them, we selected a panel of eight aptamers displaying high binding affinity for U87MG glioma cells (K_D-values ranging between 33 and 700 nM) and no or low affinity for T98G glioma cells and other human cancer cell types, including neuroblastoma, lung, and breast. Five of these aptamers showed biological activity as well, inducing time-dependent down-regulation of extracellular regulated kinase (ERK) and cyclin D1 phosphorylation, thus indicating that they may act as inhibitory ligands for critical cell surface proteins. In addition, one of the functional selected aptamers, named GL21.T, has been further characterized [23].

As a first attempt to identify its molecular target, we performed a phospho-tyrosine kinase receptor (RTK) array analysis that provided us a clear, even if preliminary, indication that GL21.T specifically binds and inhibits Axl RTK that is overexpressed in U87MG glioma cells. This result has been validated and the functional activity of the aptamer has been characterized in detail [23].

Gl21.T revealed strong inhibition of cell migration and invasion in vitro as well as tumor growth in vivo. Moreover, we demonstrated that GL21.T is also able to internalize into target cells in a receptor-dependent manner, thus resulting an interesting cargo for tissue specific internalization of therapeutic agents [24].

More recently, with the aim to discriminate between the chemo-resistant and chemo-sensitive phenotype in NSCLC, we applied a similar approach by using A549 cells in the positive selection step and the more sensitive H460 cells in the counter-selection step [16]. We selected a set of five families of 2′-F-RNA aptamers able to distinguish between the two cell types. Among these, we identified the aptamer CL4 employing phospho-RTK array analysis that binds the epidermal growth factor receptor (EGFR). We demonstrated that CL4 inhibits EGFR-mediated signaling in vitro and tumor growth in vivo.

In conclusion, the identification of the molecular signature of cancer cells is essential for an accurate and early diagnosis and the generation of molecular probes for molecular analysis represents a major goal in oncology. In this respect, differential cell-SELEX strategy offers a great promise for cancer biomarker discovery and therapeutic molecule identification.

In this chapter, we provide a detailed description of the differential cell-SELEX approach that has a general applicability to obtain a molecular signature of cell state in several cancer types.

2   Materials

2.1 In Vitro RNA Transcription

1. Transcription buffer (5×): 30 mM $MgCl_2$, 50 mM NaCl, 200 mM Tris–HCl, pH 7.5 and 10 mM spermidine.

2. To prepare transcription reaction mix add: 1× Transcription buffer, 1 mM 2′-F-2′-dCTP and 2′-F-2′-dUTP, 1 mM ATP, 1 mM GTP, 1 U/ml inorganic pyrophosphatase, 0.5 U/µl RNAse inhibitors, 10 mM dithiothreitol (DTT), 10 µCi/µl 32P-αUTP (3000 Ci/mmol), 2.5 U/µl of T7^{Y639F} RNA polymerase, and 1 pmol/µl DNA in RNAse-free water to a final volume of 600 µl.

3. DNase treatment: DNase I.

4. Denaturing polyacrylamide gel 8 %: dissolve 60 ml 40 % acrylamide–bis solution (37.5:1) and 126 g urea (*see* **Note 1**) in 30 ml of TBE10× [Tris (0.89 M)–borate (0.89 M)-EDTA (0.025 M)].

   Add water to a final volume of 300 ml. To polymerize the gel add 43 µl N,N,N',N'-tetramethyl-ethylendiamine (TEMED) and 450 µl of 10 % ammonium persulfate solution (APS) (*see* **Note 2**) for 50 ml of mix.

5. Gel loading buffer: 480 µl of formamide, 10 µl 0.5 M ethylenediamine tetraacetic acid (EDTA), 10 µl water, and bromophenol Blue.

6. Buffer to elute RNA from gel: 300 mM NaAc with 200 mM EDTA.

2.2 Counter-selection and Selection Steps

1. Buffer used for incubation during both steps: RPMI or Dulbecco's modified Eagle medium (DMEM) serum-free.
2. Washing buffer: RPMI or DMEM serum-free.
3. Nonspecific competitor: polyinosinic acid (Poly-I).
4. RNA extraction: kit used from Ambion Inc.

2.3 Reverse-Transcription Polymerase Chain Reaction (RT-PCR)

1. Reverse-transcription reaction mix: 1× buffer, 1 mM dATP, 1 mM dGTP, 1 mM dCTP, 1 mM dTTP, 1 μM reverse-primer, 100 μl RNA template (*see* **Note 3**), and 2 U M-MuLV RT.
2. PCR buffer (10×): 100 mM Tris–HCl (pH 8.3), 15 mM $MgCl_2$, 500 mM KCl.
3. PCR mix: 1× buffer, 1 mM dATP, 1 mM dGTP, 1 mM dCTP, 2 μM primers, 6 mM $MgCl_2$ (*see* **Note 4**), reverse-transcribed template, and Taq polymerase (0.02 U/μl).

2.4 TOPA-TA Cloning

1. PCR buffer (10×): 100 mM Tris–HCl (pH 8.3), 15 mM $MgCl_2$, 500 mM KCl.
2. To prepare PCR reaction mix add: PCR buffer (1×), 200 μM dATP, 200 μM dGTP, 200 μM dCTP, 200 μM dTTP, 2 μM primers (forward and reverse), DNA template from the final cycle of SELEX, and Taq polymerase (0.02 U/μl) (*see* **Note 5**).
3. TOPO-TA Cloning kit from Invitrogen.
4. TOPO cloning reaction: 4 μl PCR product, 1 μl pCR™2.1-TOPO, 1 μl Salt Solution: 200 mM NaCl, and 10 mM $MgCl_2$.
5. DH5α™ competent cells.
6. LB plates containing 50 μg/ml ampicillin or 50 μg/ml kanamycin.
7. 40 mg/ml X-gaL in dimethylformamide (DMF).

2.5 Binding Assay

1. Buffer for dephosphorylation (10×): 1 mM EDTA, 500 mM Tris–HCl, pH 8.5.
2. Enzyme for dephosphorylation: phosphatase alkaline (PA) 1 U/μl.
3. Inactivation of PA: 200 mM ethylene glycol tetraacetic acid (EGTA).
4. Buffer for phosphorylation (10×): 100 mM $MgCl_2$, 500 mM Tris–HCl, pH 8.2, 50 mM DTT, 1 mM EDTA, 1 mM spermidine.
5. Enzyme for phosphorylation: T4 Polynucleotide Kinase 10 U/μl.
6. Incubation buffer for treatment: culture medium serum-free.
7. Washing buffer: culture medium serum-free.

8. Recovering buffer: 0.6 % sodium dodecyl sulfate (SDS).

9. [γ-^{32}P]ATP (6000 Ci/mmol).

10. Nonspecific competitor: polyinosinic acid (Poly-I).

2.6 Restriction Fragment Length Polymorphism (RFLP) Analysis

1. PCR buffer (10×): 100 mM Tris–HCl, pH 8.3, 500 mM KCl, 15 mM MgCl$_2$.

2. To prepare PCR reaction mix add: PCR buffer (1×), 200 μM dATP, 200 μM dGTP, 200 μM dCTP, 200 μM dTTP, 2 μM primers (forward and reverse), 0.02 U/μl Taq polymerase, and DNA template.

3. [γ-^{32}P]ATP (3000 Ci/mmol).

4. Buffer React 1 (10×) for digestion: 500 mM Tris–HCl, pH 8.0, 500 mM NaCl, 100 mM MgCl$_2$.

5. Restriction enzymes: RsaI, AluI, HaeIII, HhaI (Invitrogen).

6. Denaturing polyacrylamide gel 6 %: dissolve 45 ml 40 % acrylamide–bis solution (37.5:1) and 126 g urea (*see* **Note 1**) in 30 ml of TBE10×. Add water to a final volume of 300 ml. To polymerize the gel add 43 μl TEMED and 450 μl of 10 % APS for 50 ml of mix.

2.7 Phospho-RTK Array Analysis

1. Block buffer: 2 ml of Array Buffer 1 (R&D Systems).

2. Incubation mix: dilute 200-300 μg of cell lysate in 1.5 ml of Array Buffer 1 (R&D Systems).

3. Washing buffer: 20 ml of 1× Wash Buffer. Wash Buffer stock solution is 25× concentrated (R&D Systems) (*see* **Note 6**). Dilute 40 ml of 25× Wash Buffer into 960 ml of deionized or distilled water.

4. Detection antibody: Anti-Phospho-Tyrosine-HRP Detection Antibody (R&D Systems) diluted 1:5000 in 1× Array Buffer 2 (*see* **Note 7**).

5. Chemi Reagent Mix: mix Chemi Reagents 1 and 2 (R&D Systems) in equal volumes within 15 min of use. 1 ml of Chemi Reagent Mix is required per membrane (*see* **Note 8**).

3 Methods

3.1 Generation of the Starting RNA Library

1. The first step is the PCR amplification of the starting library that is a high complexity DNA pool (10^{14} members) containing a 45 nt random sequence, flanked by two fixed regions for PCR amplification. The primers used are:

   P20: 5′-TCCTGTTGTGAGCCTCCTGTCGAA-3′

   P10: 5′-TAATACGACTCACTATAGGGAGACAAGAATAAA CGCTCAA-3′

2. Transcription of the amplified pool is performed at 37 °C overnight in a mix containing 10 μCi/μl 32P-αUTP (3000 Ci/mmol) and 2.5 U/μl of T7^{Y639F} RNA polymerase (*see* **Note 9**).

3. Transcribed RNA is treated with DNase I (0.2 U/μl) for 30 min at 37 °C.

4. RNA is extracted with phenol/chloroform/isoamyl alcohol (25:24:1) and precipitated with 0.3 M NaAc and 3× volumes of ice cold ethanol in the presence of 0.1 mg/ml of linear acrylamide. To precipitate RNA a centrifugation at 13,200 rpm for 30 min at 4 °C is required.

5. Obtained pellet is loaded on 8 % denaturing polyacrylamide gel.

6. After electrophoresis RNA is eluted from gel with NaAc/EDTA at 42 °C for 2 h.

7. RNA is finally resuspended in sterile water and its concentration is evaluated by absorption measurement.

3.2 Differential Cell-SELEX Protocol

The general protocol of differential cell-SELEX includes repeated cycles of five steps (Fig. 2): (1) incubation of the library with non-target cells (counter-selection step); (2) Recovering of the unbound aptamers and incubation with target cells (positive selection step); (3) partitioning of unbound oligonucleotides from those that specifically bind target cells; (4) recovery of aptamers bound to target cells; (5) reverse transcription and amplification of the nucleic acid pool enriched for specific ligands. The introduction of a counter-selection step promotes the isolation of aptamers that specifically bind to the surface of a desired cell phenotype. After reiterated steps of selection and counter-selection, the resulting oligonucleotides are cloned and sequenced. Obtained sequences are screened for conserved structural elements suggestive of potential binding domains and then tested for the ability to specifically bind to target cells.

In our protocol at each round of SELEX, one or two counter-selection steps on non-target cells preceded the positive selection step on target cells.

1. Before treatment, the 2′-F-RNA pool (800–300 pmol) is subjected to a short denaturation–renaturation step in 1.5 ml of serum-free medium (85 °C for 5 min, snap-cooled on ice for 2 min and allowed to warm up to 37 °C).

2. Following the denaturation–renaturation step, 13.5 ml of serum-free medium are added to the RNA to have a final volume of 15 ml for treatment of cells plated in 150 mm dishes.

3. The 2′-F-RNA pool is first incubated at 37 °C for 30 min with non-target cells (counter-selection step).

4. Unbound sequences are then recovered and incubated at 37 °C for 30 or 15 min with target cells (selection step).

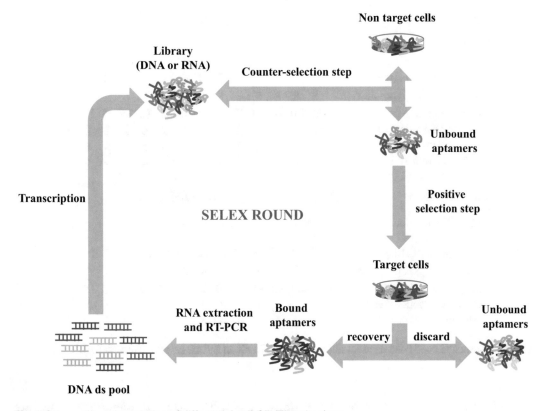

Fig. 2 Schematic representation of differential cell-SELEX technology

5. Finally, following several washings (up to five) with 5 ml of serum-free medium (to discard unbound sequences), bound aptamers are recovered by total RNA extraction.

   During the selection procedure, changing the number of final washings, the incubation time on target cells and the number of counter-selection steps progressively increases the selection-pressure. Exact SELEX conditions used for NSCLC and glioma are shown in Fig. 3.

6. To monitor the enrichment of the selected RNA pool we performed RFLP analysis and observed the occurrence of four-base restriction sites in the population, indicating the presence of distinct aptamer families.

3.3 RFLP Analysis

1. The PCR product (about 500 ng) is first end-labeled with [γ-^{32}P] ATP.

2. Labeled PCR product is digested with a mix of four restriction enzymes (RsaI, AluI, HaeIII, HhaI) in the buffer ReactI for 1 h at 37 °C.

3. Digested samples are loaded onto 6 % denaturing polyacrylamide gel and analyzed by autoradiography.

a   **NSCLC**

SELEX condition

Cell number (x10^6)	2	2	2	2	2	2	2	2	2	2	2	2	2	2
RNA amount (x100pmol)	8	8	8	8	8	8	8	7	7	7	8	7	7	6
Incubation time (min)	30	30	30	30	30	15	15	15	15	15	15	15	15	15
Counter-selection number	1	1	1	1	2	2	2	2	2	2	2	2	2	2
Washing number	3	3	3	3	3	3	3	3	3	5	5	5	5	5
PolyI (10ng/ul)	-	-	-	-	-	-	-	-	-	-	-	-	-	+ +

b   **GLIOMA**

SELEX condition

Cell number (x10^7)	10	10	10	10	10	10	10	10	10	10	10	10	10	10
RNA amount (x100pmol)	5	8	8	3	3	3	3	3	3	3	3	3	3	3
Incubation time (min)	30	30	30	30	30	30	30	30	30	15	15	15	15	15
Counter-selection number	1	1	1	1	1	1	1	1	1	1	2	2	2	2
Washing number	1	2	3	3	4	4	5	5	5	5	5	5	5	5

Fig. 3 Experimental conditions used for differential cell-SELEX on NSCLC (**a**) and glioma cells (**b**)

3.4 Cloning and Sequencing of Selected Aptamers

1. After 14 rounds of SELEX, selected oligonucleotides are cloned in TOPO-TA Cloning Kit according to the manufacturer's protocol and about 100 clones are sequenced.

2. Obtained sequences are analyzed by using bioinformatics alignment tools to identify and score sequence patterns present in aptamers. This approach allows to recognize conserved and variable regions and to regroup sequences into quasi-phylogenetic families. Usually, conserved motifs are indicative of specific target recognition domains.

3.5 Binding Assay

Binding of the final pool, individual aptamers or starting pool (used as a control) to target cells is performed in 24-well plates in triplicate.

1. Aptamers are first dephosphorylated with PA at 37 °C for 1 h.

2. Following dephosphorylation, aptamers are end-labeled at 5' by using T4 polynucleotide kinase in the presence of [γ-^{32}P] ATP at 37 °C for 30 min.

3. 3.5×10^4 cells per well are incubated with various concentrations of pools or individual aptamers in 200 μl of serum-free medium for 20 min at room temperature in the presence of 0.1 mg/ml Poly-I as a nonspecific competitor.

4. After five washings with 500 μl of serum-free medium, bound sequences are recovered with 300 μl of SDS 0.6 % and the amount of radioactivity is measured at the beta counter.

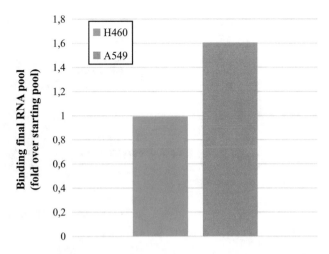

Fig. 4 Binding analyses of the final RNA pool to H460 (non-target) and A549 (target) NSCLC cells. RNA was 5′-[^{32}P]-labeled and incubated at 50 nM with cells. The results are expressed as fold over the binding detected with the starting pool

5. The background value obtained with the starting pool is subtracted from the value obtained with the specific aptamers or final pools.

6. K_D values for each aptamer are determined by Linewaver Burk analysis according to the equation:

$$1/[\text{complex}] = K_D/[\text{Cmax}] \times 1/[\text{aptamer}] + 1/[\text{Cmax}].$$

An example of the binding result obtained with final pool following differential cell-SELEX on NSCLC is reported in Fig. 4.

3.6 Target Identification

Since the binding of an aptamer to cell surface receptors often results in the inhibition of the target receptor and its intracellular signaling, to identify the putative functional targets of aptamers with the best binding properties one possibility is to perform a phospho-RTK array analysis.

1. Target cells are serum starved overnight, pretreated with 200 nmol/l aptamer for 3 h and then stimulated with 20 % FBS either alone or in presence of the aptamer.

2. Cell lysates are incubated overnight with the phospho-RTK array and, following three washings with 1× Wash buffer, incubated with anti-Phospho-Tyrosine-HRP Detection Antibody for 2 h.

3. Chemiluminescent signal is detected using Chemi Reagent Mix.

This approach appears to be very useful and convenient as it is informative but very easy and fast to perform.

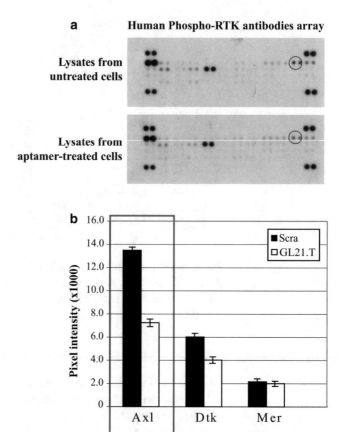

Fig. 5 (**a**) Lysates from untreated (*upper panel*) or GL21.T treated (*lower panel*) U87MG glioma cells were incubated with Human Phospho-RTK array (R&D Systems). The *red circles* indicate the signal coming from hybridization with an anti-Axl. Intensity of signals was measured for Axl and other two members (Dtk and Mer) belonging to the same receptor family. Pixel intensity is reported in the graph (**b**). A clear reduction of Axl phosphorylation was observed. Figure partially reproduced from Ref. [23]

A representative result obtained for the GL21.T aptamer by phospho-RTK array analysis is shown in Fig. 5 (adapted from Ref. [23]). Gl21.T aptamer shows to inhibit Axl phosphorylation. This circumstance has been further validated in our laboratory by different approaches [23].

4   Notes

1. Dissolve urea by adding it slowly. If urea precipitates in the denaturing polyacrylamide solution, warm it to 37 °C to redissolve prior to prepare the gel.

2. Prepare 10 % APS solution in water and immediately freeze it in aliquots at −20 °C.

3. Once extracted, RNA must be resuspended in 100 μl sterile water.

4. The concentration of $MgCl_2$ takes into account that 1× PCR buffer contains 1.5 mM $MgCl_2$. The total final concentration of $MgCl_2$ must be 7.5 mM.

5. It is important to include a 15 min of extension at 72 °C after the last cycle to ensure that all PCR products are full length and 3′-adenylated.

6. If crystals have formed in the concentrated solution, warm the bottle to room temperature and mix gently until the crystals have completely dissolved.

7. To obtain 1× Array Buffer 2, add 2 ml of Array Buffer 2 Concentrate 5× to 8 ml of deionized or distilled water.

8. Protect Chemi Reagents 1 and 2 and Chemi Reagents Mix from light.

9. Mutant T7Y^{639F} RNA polymerase has the ability to efficiently incorporate 2′-modified-ribonucleotides. 2′F- modification of the RNAs increases resistance to nucleases.

Acknowledgment

This work was supported by funds from: MIUR grant, MERIT RBNE08YFN3_001 (VdF), AIRC # 13345 (VdF); from the Italian Ministry of Economy and Finance to the CNR for the Project FaReBio di Qualità (VdF); Grant CNR "Medicina Personalizzata" (VdF); Compagnia San Paolo # 2011.1172 (VdF); CNR Flagship Project NanoMax (*DESIRED*) 2012–2014 (VdF). Figure 5 was adapted from Ref. [23] by courtesy of Mol. Ther. (Nature, London)

References

1. Cerchia L, de Franciscis V (2010) Targeting cancer cells with nucleic acid aptamers. Trends Biotechnol 28:517–525

2. Missailidis S, Hardy A (2009) Aptamers as inhibitors of target proteins. Expert Opin Ther Pat 19:1073–1082

3. Keefe AD, Pai S, Ellington A (2010) Aptamers as therapeutics. Nat Rev Drug Discov 9:660

4. Ferreira CS, Matthews CS, Missailidis S (2006) DNA aptamers that bind to MUC1 tumour marker: design and characterization of MUC1-binding single-stranded DNA aptamers. Tumor Biol 27:289–301

5. Lupold SE, Hicke BJ, Lin Y et al (2002) Identification and characterization of nuclease-stabilized RNA molecules that bind human prostate cancer cells via the prostate-specific membrane antigen. Cancer Res 62:4029–4033

6. Chen CH, Chernis GA, Hoang VQ et al (2003) Inhibition of heregulin signaling by an aptamer that preferentially binds to the oligomeric form of human epidermal growth factor receptor-3. Proc Natl Acad Sci U S A 100:9226–9231

7. Li N, Larson T, Nguyen HH et al (2010) Directed evolution of gold nanoparticle delivery to cells. Chem Commun (Camb) 46:392–394

8. Ohuchi SP, Ohtsu T, Nakamura Y (2006) Selection of RNA aptamers against recombinant

transforming growth factor beta type III receptor displayed on cell surface. Biochimie 88:897–904

9. Chen F, Hu Y, Li D et al (2009) CS-SELEX generates high-affinity ssDNA aptamers as molecular probes for hepatitis C virus envelope glycoprotein E2. PLoS One 4:e8142

10. Shangguan D, Li Y, Tang Z et al (2006) Aptamers evolved from live cells as effective molecular probes for cancer study. Proc Natl Acad Sci U S A 103:11838–11843

11. Chen HW, Medley CD, Sefah K et al (2008) Molecular recognition of small-cell lung cancer cells using aptamers. ChemMedChem 3:991–1001

12. Sefah K, Meng L, Lopez-Colon D et al (2010) DNA aptamers as molecular probes for colorectal cancer study. PLoS One 5:e14269

13. Thiel KW, Hernandez LI, Dassie JP et al (2012) Delivery of chemo-sensitizing siRNAs to HER2+-breast cancer cells using RNA aptamers. Nucleic Acids Res 40:6319–6337

14. Thiel WH, Bair T, Peek AS et al (2012) Rapid identification of cell-specific, internalizing RNA aptamers with bioinformatics analyses of a cell-based aptamer selection. PLoS One 7:e43836

15. Cerchia L, Esposito CL, Jacobs AH et al (2009) Differential SELEX in human glioma cell lines. PLoS One 4:e7971

16. Esposito CL, Passaro D, Longobardo I et al (2011) A neutralizing RNA aptamer against EGFR causes selective apoptotic cell death. PLoS One 6:e24071

17. Louis DN, Ohgaki H, Wiestler OD et al (2007) The 2007 WHO classification of tumours of the central nervous system. Acta Neuropathol 114:97–109D

18. Louis DN (2006) Molecular pathology of malignant gliomas. Annu Rev Pathol 1:97–117

19. Mason WP, Cairncross JG (2008) Invited article: the expanding impact of molecular biology on the diagnosis and treatment of gliomas. Neurology 71:365–373

20. Rao RD, Uhm JH, Krishnan S et al (2003) Genetic and signaling pathway alterations in glioblastoma: relevance to novel targeted therapies. Front Biosci 8:e270–e280

21. Sathornsumetee S, Rich JN (2008) Designer therapies for glioblastoma multiforme. Ann N Y Acad Sci 1142:108–132

22. Wen PY, Kesari S (2008) Malignant gliomas in adults. N Engl J Med 359:877

23. Cerchia L, Esposito CL, Camorani S et al (2012) Targeting Axl with an high-affinity inhibitory aptamer. Mol Ther 20:2291–2303

24. Esposito CL, Cerchia L, Catuogno S et al (2014) Multifunctional aptamer-miRNA conjugates for targeted cancer therapy. Mol Ther 22:1151–1163

Chapter 4

DNA Aptamer Generation by Genetic Alphabet Expansion SELEX (ExSELEX) Using an Unnatural Base Pair System

Michiko Kimoto, Ken-ichiro Matsunaga, and Ichiro Hirao

Abstract

Genetic alphabet expansion of DNA using unnatural base pair systems is expected to provide a wide variety of novel tools and methods. Recent rapid progress in this area has enabled the creation of several types of unnatural base pairs that function as a third base pair in polymerase reactions. Presently, a major topic is whether the genetic alphabet expansion system actually increases nucleic acid functionalities. We recently applied our unnatural base pair system to in vitro selection (SELEX), using a DNA library containing four natural bases and an unnatural base, and succeeded in the generation of high-affinity DNA aptamers that specifically bind to target proteins. Only a few hydrophobic unnatural bases greatly augmented the affinity of the aptamers. Here, we describe a new approach (genetic alphabet *Expansion SELEX*, ExSELEX), using our hydrophobic unnatural base pair system for high affinity DNA aptamer generation.

Key words Genetic alphabet expansion, Unnatural base pair, Aptamer, SELEX, In vitro selection, PCR

1 Introduction

Replication and transcription of nucleic acids rely on two sets of base pairs, A–T and G–C, and thus the functions of nucleic acid molecules, such as aptamers that specifically bind to target molecules, generated by in vitro selection (SELEX) involving PCR amplification [1, 2], are restricted within this limited number of only four natural base components. To further increase the affinity and stability of nucleic acid aptamers and the success rates of aptamer generation, several modified components based on the natural bases have been applied to SELEX [3–5]. However, these natural-base modification methods have not yet substantially improved the affinities of aptamers with high specificity to target proteins.

Günter Mayer (ed.), *Nucleic Acid Aptamers: Selection, Characterization, and Application*, Methods in Molecular Biology, vol. 1380, DOI 10.1007/978-1-4939-3197-2_4, © Springer Science+Business Media New York 2016

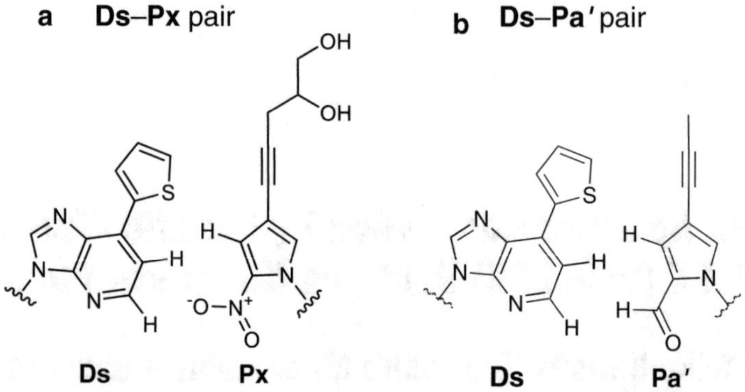

Fig. 1 Chemical structures of the Ds–Px and Ds–Pa′ pairs. (**a**) The Ds–Px pair functions as a third base pair in PCR with high efficiency and fidelity, by using both the Ds and Px substrates. (**b**) The Ds–Pa′ pair is employed for replacement PCR, by using just the Pa′ substrate, without dDsTP

Another approach toward the substantial improvement of SELEX might be the expansion of the genetic alphabet of nucleic acids. As compared to antibodies composed of 20 different amino acids, standard nucleic acids consist of only four nucleotide bases with similar chemical and physical properties. Thus, increasing the number of nucleotides by creating artificial extra bases (unnatural bases) with different properties had been expected to augment nucleic acid functionality, such as aptamer affinities. As a requirement for use in the genetic alphabet expansion in SELEX, an unnatural base should function as a third base pair with its counterpart in PCR amplification.

So far, three types of unnatural base pairs have been created for conventional PCR amplification [6–10]. Among them, the hydrophobic base pair between 7-(2-thienyl)imidazo[4,5-*b*]pyridine (Ds) and 2-nitro-4-propynylpyrrole (Px) that we developed exhibits high fidelity and efficiency in PCR, by either DeepVent or AccuPrime*Pfx* DNA polymerase (Fig. 1a) [7, 8].

Recently, we developed a new SELEX method using the Ds–Px pair (genetic alphabet *Ex*pansion *SELEX*, ExSELEX), for generating DNA aptamers from a library containing hydrophobic Ds bases (Fig. 2) [11]. We added only the Ds base as the fifth base to the library for two reasons: (1) the hydrophobic Ds base strengthens the interactions with hydrophobic portions in target proteins (increasing the chemical diversity), and (2) in the absence of the pairing partner Px bases in the library, the Ds bases cannot pair in the tertiary structures of each DNA fragment (increasing the structural diversity).

A major problem in developing ExSELEX is the difficulty in determining the sequence of each aptamer in the isolated DNA library, after finishing several rounds of selection and PCR amplifi-

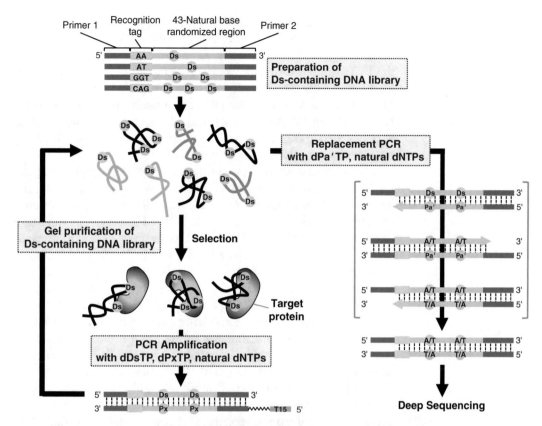

Fig. 2 Scheme of genetic alphabet expansion SELEX (ExSELEX) for generating DNA aptamers from a library containing hydrophobic Ds bases. The steps described in this chapter are highlighted in *boxes*

cation. In conventional SELEX, each aptamer sequence can be determined by cloning and sequencing, or deep sequencing using a next-generation sequencer. However, at present, we have no means of utilizing these methods and sequencers involving unnatural base pairs. To address this difficulty, we developed a new method for ExSELEX, using a DNA library containing more than 20 different sublibraries (Fig. 2) [11].

Each sublibrary consists of a 43-natural-base randomized region, in which one to three Ds bases are embedded at different predetermined positions in each sublibrary, a recognition tag sequence with two or three natural bases, which is specific to each sublibrary, and PCR primer sequences at both termini. Each sublibrary is chemically synthesized and mixed to prepare the ExSELEX DNA library.

In ExSELEX, conventional methods are employed for the selection step of isolating DNA fragments that bind to target proteins from the library. After selection, the isolated DNA fragments are amplified by PCR using AccuPrime*Pfx* DNA polymerase with the four natural base substrates and two unnatural base substrates, Ds and Px. For PCR in ExSELEX, AccuPrime *Pfx* DNA polymerase is better than DeepVent DNA polymerase. Although the fidelity of the Ds–Px pair using DeepVent DNA polymerase is slightly higher than that using AccuPrime *Pfx* DNA polymerase, the PCR amplification efficiency and sequence independence using AccuPrime *Pfx* DNA polymerase are superior to those using DeepVent DNA polymerase [8]. To separate the single-stranded Ds-containing DNA fragments from the amplified duplexes, PCR is performed using a primer linked with an extra oligo-T (T15) sequence, via a C12 alkyl spacer.

After several rounds of selection and PCR amplification, the Ds bases in the selected DNA library are replaced with natural bases (mainly adenine bases) by PCR amplification in the absence of dDsTP and dPxTP (replacement PCR). However, the mutation of the unnatural to natural bases by replacement PCR is inefficient, due to the high fidelity of the Ds–Px pair in PCR. Thus, another unnatural base substrate, a nucleoside triphosphate of 4-propynylpyrrole-2-carbaldehyde (Pa′) [6, 12], is added as a mediator from the unnatural to natural base replacement in PCR. The fidelity of Pa′ as a pairing partner of Ds is lower than that of the Px base, and thus the misincorporation rate of natural base substrates opposite Pa′ is increased after the Pa′-incorporation into DNA opposite Ds (Fig. 1b). For replacement PCR, we use TITANIUM Taq or AccuPrime *Pfx* DNA polymerase. After replacement PCR, each sequence consisting of natural bases is determined by deep sequencing, and the original Ds positions in each sequence are identified from the recognition tag sequence. From the determined sequences, Ds-containing DNA aptamers are chemically synthesized, and the affinity of each DNA aptamer is evaluated by binding assays with the target proteins [11].

Here, we describe the several steps focused on the preparation of the Ds-containing DNA library, the PCR amplification, and the replacement PCR in ExSELEX, since any conventional processes can be chosen, depending on the target proteins, for the other steps (selection method and sequencing determination) [13–18]. However, the unnatural base pair technology is quite new, and the present ExSELEX method may be modified according to future improvements in the unnatural base pair systems.

2 Materials

The common materials used in the experiments are listed below (*see* **Note 1**).

1. Apparatus for gel electrophoresis (*see* **Note 2**).

2. Nuclease-free, deionized, sterile water.

3. 40 % acrylamide–bis (19:1) solution (*see* **Note 3**).

4. Nuclease-free 10× TBE: 0.89 M Tris base, 0.89 M boric acid, and 0.02 M EDTA (*see* **Note 2**).

5. Urea (ultrapure grade).

6. Acrylamide–bis solution with urea in 1× TBE (*see* **Note 4**).

7. APS: 10 % ammonium persulfate (w/v). This can be stored at 4 °C for 1–2 months (*see* **Note 5**).

8. ≥99 % TEMED: N,N,N',N'-tetramethylethylenediamine (*see* **Note 5**).

9. Polyacrylamide gel containing 7 M urea (*see* **Note 5**).

10. TLC plate containing a UV fluorescent indicator (*see* **Note 6**).

11. Handheld UV lamp (254 nm).

12. Saran Wrap.

13. Razor blade.

14. Filter units: Ultrafree-CL and Steriflip (pore size 0.22 μm, Millipore) (*see* **Note 7**).

15. 3 M sodium acetate, pH 5.2 (*see* **Note 8**).

16. 99.5 % ethanol (*see* **Note 8**).

17. Denaturing gel-loading buffer: 10 M urea solution in 1× TBE (*see* **Note 9**).

18. 2 mM each dNTP mix: Mix equal amounts of standard deoxyribonucleoside triphosphates of the natural bases (dATP, dTTP, dCTP, and dGTP), and dilute into a 2 mM working solution. Aliquots of the solution should be stored at –10 to –30 °C.

19. DNA ladder marker (e.g., 10, 25, or 100 bp ladder).

20. 10 mg/ml ethidium bromide (*see* **Notes 2** and **10**).

21. 4 % agarose gel: To prepare a 4 % agarose gel, add 4 g of NuSieve 3:1 Agarose (Lonza) to 100 ml of 1× TBE, and heat until the agarose is completely dissolved. Add 10 mg/ml ethidium bromide at the final proportion of 0.1 μl/ml, pour the gel in a gel-casting chamber, insert the comb and let cool at room temperature for at least 30 min.

22. Agarose-loading buffer: 30 % glycerol, 0.025 % (w/v) BPB.

2.1 Preparation of Ds-Containing DNA Library

1. Chemically synthesized Ds-containing DNA sublibraries (*see* **Note 11**): The DNA library with the sequence, 5′-CTGT CAATCGATCGTATCAGTCCAC-Y2₋₃-N₄₃-GCATGACTCG AACGGATTAGTGACTAC-3′, is designated as N43Ds and used in our aptamer selection (*see* **Note 12**) [11]. Y_{2-3} is a recognition tag sequence with two or three natural bases. N_{43} is a randomized sequence with 43 natural bases, in which one to three Ds bases are embedded at different predetermined positions in each sublibrary. Underlined sequences are constant

regions for annealing sites with forward and reverse primers in PCR amplification.

2. 8 % polyacrylamide gel, 7 M urea (20 cm×40 cm, 2 mm thickness) (*see* **Notes 2**, **4**, and **5**).

2.2 Preparation of Ds-Containing DNA Library by PCR Amplification

1. Ds-containing DNA library.

2. Primers: For the PCR amplification of the N43Ds library, the following two primers can be used: Fwd27 primer, 5′-TTCTGTCAATCGATCGTATCAGTCCAC-3′, and T15-L-Rev29 primer, 5′-TTTTTTTTTTTTTTT-C12spacer-AAGTAGTCACTAATCCGTTCGAGTCATGC-3′ (*see* **Note 13**) [11].

3. 2.5 U/μl AccuPrime *Pfx* DNA polymerase (Life Technologies): The components are AccuPrime *Pfx* DNA polymerase (2.5 U/μl), 50 mM MgSO₄, and 10× AccuPrime *Pfx* Reaction Mix, which includes final concentrations of 1 mM MgSO₄ and 0.3 mM each natural dNTP. In PCR amplification involving the Ds–Px pair, we additionally supplement the reaction mixture with 0.1 mM each natural dNTP, 0.5 mM MgSO₄, and 0.05 mM each of dDsTP and Diol1-dPxTP (*see* **Note 14**).

4. 0.5 mM dDsTP (TagCyx Biotechnologies) (*see* **Note 15**) [6].

5. 0.5 mM Diol1-dPxTP (TagCyx Biotechnologies) (*see* **Note 15**) [8].

6. 10 % polyacrylamide gel, 7 M urea (16 cm×16 cm, 2 mm thickness) (*see* **Notes 2**, **4**, and **5**).

2.3 Replacement PCR

1. Ds-containing DNA library.

2. Primers: Fwd27 primer, 5′-TTCTGTCAATCGATCGTATCAGTCCAC-3′, and Rev29 primer, 5′-AAGTAGTCACTAATCCGTTCGAGTCATGC-3′ [11].

3. 10× TITANIUM Taq PCR Buffer (Clontech).

4. 50× TITANIUM Taq DNA polymerase (Clontech).

5. 0.25 mM dPa′TP (TagCyx Biotechnologies) (*see* **Note 15**) [6, 12].

6. PCR product purification kit: e.g., Wizard® SV Gel and PCR Clean-Up System, MinElute PCR Purification Kit (QIAGEN).

3 Methods

3.1 Preparation of Ds-Containing DNA Library

1. Set up gel electrophoresis to purify each sublibrary (*see* **Note 2**): Attach the gel plate to the electrophoresis apparatus, fill the upper and lower reservoirs with 1× TBE as a running buffer, and connect to an electric power supply. Prior to loading the sample, run for 30–60 min until the temperature of the gel reaches around 50 °C.

2. Prepare the loading sample: Dissolve DNA fragments in 200 μl of water and mix with 200 μl of denaturing gel-loading buffer. Heat the solution at 75 °C for 3 min.

3. Turn off the power supply, flush out the urea leached from the wells with 1× TBE, and then load the sample in each well.

4. Electrophorese until the marker dyes reach the predetermined positions. Generally, stop the electrophoresis when the xylene cyanol dye reaches about one-fourth of the distance from the bottom of the gel.

5. After electrophoresis, remove the gel plate from the apparatus. Detach one glass plate, and cover the gel with Saran Wrap. Turn the glass plate over, and remove the other glass plate from the gel. Place the gel on a fluorescent TLC plate. Check the shadow band corresponding to the DNA fragments on the gel by illumination with a handheld UV lamp at 254 nm (*see* **Note 6**).

6. Excise the bands with a clean razor blade, transfer the gel slices to a 50-ml conical tube, and crush the slices (*see* **Note 16**). Add 5.4 ml of sterile water, and incubate the tube for about 10 h at 37 °C with gentle agitation.

7. Pass the eluted solution through a 0.22 μm Steriflip filter (*see* **Note 7**).

8. Place 1.2 ml aliquots of the filtrate (about 4.8 ml) into four 5 ml tubes. Add 120 μl of 3 M sodium acetate, and 2.5 ml of ethanol per tube. After mixing, store them at –20 °C for 1 h. Recover the precipitated DNA fragments by centrifugation (*see* **Note 8**).

9. Add 0.1 ml water to each tube and incubate for 3 min at 75 °C, to dissolve the purified DNA fragments.

10. Combine the solutions in a single 1.5 ml tube, and determine the oligonucleotide concentration from its UV absorbance in an appropriate buffer, such as TE buffer.

11. Make the Ds-containing DNA library: Mix equal amounts of each sublibrary, and dilute to a 10 μM working solution. Store the solution in aliquots at –10 to –30 °C.

3.2 Preparation of Ds-Containing DNA Library by PCR Amplification

1. Prepare a master mix in a 1.5 ml tube, for 9-tube PCR reactions on a 0.1 ml scale (*see* **Note 17**). *See* Table 1.

2. Transfer a 60 μl volume from the master mix to a thin-wall PCR tube, add 40 μl of water and mix the solution (designated as Non-Template-Control, "NTC").

3. To the remaining master mix (480 μl), add 320 μl of Ds-containing DNA library and mix the solution, and then aliquot 100 μl volumes from the solution into eight thin-wall PCR tubes (designated as "DNA plus").

Table 1
Protocol for setting up the PCR reaction using the modified nucleotide dDsTP

Component	Volume (µl)	Final concentration in PCR
10× AccuPrime *Pfx* Reaction Mix	90	1×
20 µM Fwd27 primer	45	1 µM
20 µM T15-L-Rev29 primer	45	1 µM
2 mM each dNTP mix	45	0.4 mM in total
10 mM MgSO$_4$	45	1.5 mM in total
0.5 mM dDsTP	90	0.05 mM
0.5 mM Diol1-dPxTP	90	0.05 mM
Water	72	
2.5 U/µl AccuPrime *Pfx* DNA polymerase	18	0.5 U/µl

4. Place the "DNA plus" and "NTC" reaction tubes in a thermal cycler, and perform PCR with a program consisting of *N* cycles at 94 °C for 30 s and 65 °C for 2 min 30 s (*see* **Note 17**).

5. Combine the "DNA plus" PCR reactions into a single 1.5-ml tube.

6. Mix 5 µl each of "DNA plus" and "NTC" reaction with 2 µl agarose-loading buffer and load the samples onto a 4 % agarose gel. Load a DNA ladder marker in one lane.

7. Run the gel at 150 V with 1× TBE as the running buffer for 15 min, and detect the dsDNA bands with a UV-lamp.

8. Set up the gel electrophoresis to purify the PCR-amplified, Ds-containing DNA library (*see* **Note 2**): Attach the gel plate to the electrophoresis apparatus, fill the upper and lower reservoirs with 1× TBE as a running buffer, and connect to an electric power supply. Prior to loading the sample, run the gel for about 30 min until the temperature of the gel reaches around 50 °C.

9. Prepare the loading sample: Add 400 µl of denaturing sample solution to the "DNA plus" PCR samples (total about 800 µl, prepared in **step 4**). Heat the solution at 75 °C for 3 min.

10. Turn off the power supply, flush out the urea leached from the wells with 1× TBE, and then load the sample inside three large wells (400 µl × 3) (*see* **Note 18**).

11. Run the gel until the marker dyes reach the predetermined positions. Generally, stop the electrophoresis run when the bromophenol blue dye has almost reached the bottom of the gel.

12. After gel electrophoresis, remove the gel plate from the apparatus. Detach one glass plate, and cover the gel with Saran Wrap. Turn the glass plate over, and remove the other glass plate from the gel. Place the gel on a fluorescent TLC plate. Check the shadow bands corresponding to the DNA fragments on the gel (*see* **Note 19**), by illumination with a hand-held UV lamp at 254 nm (*see* **Note 6**).

13. Excise the band corresponding to the Ds-containing DNA library (*see* **Note 19**), with a clean, disposable razor blade, transfer each gel slice to three 2-ml tubes, and crush the slices against the wall of the tube (*see* **Note 16**). Add 1 ml of water to each tube, and incubate the tube for 10 h at 37 °C with gentle agitation.

14. Pass the eluted solution through a 0.22-μm Ultrafree-CL filter unit (*see* **Note 7**).

15. Mix the filtrate with 1/10 volume of 3 M sodium acetate, and 2 volume of ethanol. After mixing, store at –20 °C for 1 h. Recover the precipitated DNA fragments by centrifugation (*see* **Note 8**).

16. Dissolve the purified DNA library in 50–200 μl of water. Determine the DNA concentration from its UV absorbance.

3.3 Replacement PCR

1. Prepare the PCR reaction mix in a thin-wall tube. *See* Table 2 for a 100-μl PCR reaction.

2. Perform the PCR reactions in a thermal cycler with the following PCR conditions: 10–12 cycles at 94 °C for 30 s and 65 °C for 2 min 30 s (*see* **Note 20**).

3. Mix 5 μl of the PCR sample with 2 μl agarose-loading buffer, and load the sample on a 4 % agarose gel. Load a DNA ladder maker in one lane.

4. Run the gel at 150 V with 1× TBE as a running buffer for 15 min, and detect the dsDNA bands with a UV-lamp.

5. Purify the PCR products by using PCR product purification kits, to remove unconsumed primers, dNTPs, and other materials.

6. Determine the DNA concentrations from the UV absorbance and subject the samples to sequencing procedures, such as cloning and preparation of DNA templates for deep sequencing.

Table 2
Protocol for setting up the PCR reaction using the modified nucleotide dPa′TP

Component	Volume (μl)	Final concentration
10× TITANIUM Taq PCR Buffer	10	1×
20 μM Fwd27 primer	5	1 μM
20 μM Rev29 primer	5	1 μM
2 mM each dNTP mix	15	0.3 mM each
0.25 mM dPa′TP	20	0.05 mM
Water	3	
50× TITANIUM Taq PCR Buffer	2	1×
5 nM Ds-containing DNA library	40	2 nM (*see* **Note 20**)

4   Notes

1. It is important to take care to avoid nucleases and DNA contamination from conceivable sources. Use gloves during experiments, and DNA AWAY (Molecular BioProducts, Inc.) is effective to remove DNA contamination on laboratory instruments and glass/plastic-ware.

2. To analyze small PCR products (less than 100 bp), we use a commercially supplied agarose gel electrophoresis system or a vertical slab electrophoresis system with a denaturing polyacrylamide gel (glass plate size: 16 cm × 16 cm, thickness: 1 mm). For rapid confirmation of PCR products with no need for purification and quantification, we analyze the products on a 4 % agarose gel with ethidium bromide, using 1× TBE as the running buffer. For purification and quantification of the products, we generally use a denaturing polyacrylamide gel. For the purification of chemically synthesized DNA fragments (0.2-μmol scale synthesis), we use a denaturing gel electrophoresis system with a longer plate size (20 cm × 40 cm, 2 mm thickness), because higher resolution is needed to separate a one-base difference in oligonucleotides. A power supply with a temperature probe is useful for automatic and precise temperature control during the denaturing gel electrophoresis. We usually set the temperature at 45–50 °C and use 1× TBE as the running buffer.

3. Monomeric acrylamide is neurotoxic, and care should be taken to avoid exposure. Store the solution at 4 °C with protection from light.

4. To make denaturing polyacrylamide gels, it is useful to have a stock solution, which is kept at 4 °C. For example, to make a 10 % acrylamide–bis solution with 7 M urea (400 ml), mix 100 ml of 40 % acrylamide–bis (19:1) solution, 40 ml of 10× TBE, and 168 g of urea, and bring the volume to 400 ml with water. To remove impurities, filtration is recommended, but not required. The optimal acrylamide–bis percentage depends on the nucleotide length to be separated. The percentage of polyacrylamide in the gel should be lower with longer oligonucleotides. In general, for >90-mer oligonucleotides, an 8 % denaturing polyacrylamide is used, with 10 % for 45–120-mer, 15 % for 25–50-mer, and 20 % for <30-mer.

5. For example, to prepare a 20 cm×40 cm (2 mm) gel, mix 130 ml acrylamide–bis solution with 650 µl of APS. Just before pouring, add 130 µl of TEMED, and mix the solution by gentle swirling. Immediately pour the gel mix between the gel plates and insert the gel comb. For oligonucleotide purification, making the gel on the previous day is recommended to ensure complete polymerization, although the gel usually polymerizes in about 30 min. After removing the comb, promptly wash out the wells thoroughly. To prepare a 16 cm×16 cm (2 mm) gel, 40 ml acrylamide–bis solution, 200 µl of APS, and 40 µl of TEMED are used.

6. Highly concentrated oligonucleotides (>0.1 OD units) within a gel can be visualized by ultraviolet (UV) shadowing, without staining. Place the gel, covered with Saran Wrap, onto the TLC plate containing a fluorescent indicator, and shine UV light from a handheld UV lamp. The regions with highly concentrated nucleotides can be detected as "shadows" on the plate, due to their absorbance of the illuminated UV.

7. Before ethanol precipitation of the purified oligonucleotides eluted from the polyacrylamide gel slices, the solution should be filtered through a filter unit, to remove the small gel slices. Otherwise, these impurities will also be precipitated, and not only inhibit the concentration determination by UV absorbance, but also impede further applications.

8. For ethanol precipitation, mix the solution containing DNA with 1/10 vol. of 3 M sodium acetate and 2–3 vol. of ethanol, and store the solution at −10 to −30 °C (for at least 30 min). After centrifugation (10,000–12,000×g for 40 min) at 4 °C, wash the precipitate with 500 µl of pre-chilled 70 % ethanol per tube. The ethanol residue is evaporated with a centrifugal evaporator. In the ethanol precipitation of DNA libraries used for binding with the target protein, co-precipitating agents, such as glycogen, should not be used, since such agents might affect the interactions of the DNA library with the target.

9. The solution should be filtered and stored in aliquots at −10 to −30 °C. Before use, incubate the aliquot at 75 °C to completely dissolve the urea. Alternatively, a denaturing solution with a dye, such as 0.05 % (w/v) xylene cyanol or 0.05 % (w/v) bromophenol blue as a marker dye, can also be used. In that case, keep in mind that the migration of the dye should not correspond with that of the desired fragments on the gel.

10. Ethidium bromide intercalates in DNA and is highly toxic, and care should be taken to avoid exposure.

11. Ds-containing DNA fragments can be purchased from custom suppliers, or chemically synthesized with an automated DNA synthesizer using dDs-amidite (dDs-CE Phosphoramidite, Cat. Nos: 10-1521-90 for 100 μmol, and 10-1521-02 for 0.25 g, Glen Research). In general, we perform the chemical synthesis on the 0.2 μmol scale by the conventional phosphoramidite method. The protecting group removal in the synthetic reactions (i.e., deprotection) is usually performed by heating the preparation at 55 °C for 6 h, after the fragment release from the CPG column by an incubation at room temperature for an hour in a concentrated NH_4OH solution. The NH_4OH is evaporated to dryness with a centrifugal evaporator. Afterwards, the fragments are subjected to purification by denaturing gel electrophoresis.

12. Keep in mind that the location of the Ds bases might affect the PCR amplification efficiency. In general, we separate two Ds bases by at least six natural bases when designing Ds-containing sublibraries [7, 11]. For more efficient PCR amplification with reduced sequence bias, the separation of two Ds bases by at least eight natural bases is recommended.

13. The T15-L-Rev29 primer is the Rev29 primer linked with an extra oligo-T (T15) sequence via the C12 alkyl spacer. In PCR products, the Ds-containing strand sequence is 5′-TTCTGTCAATCGATCGTATCAGTCCAC-Y2.3-N43-GCATGACTCGAACGGATTAGTGACTACTT-3′ and the Px-containing strand sequence is composed of 5′-TTTTT TTTTTTTTT-C12-spacer and the sequence complementary to the Ds-containing strand.

14. The amplification efficiency and fidelity of the Ds–Px pairing during PCR depend on the DNA polymerases used in the reactions [8]. Among the commercially available DNA polymerases, Deep Vent DNA polymerase with $3′ → 5′$ exonuclease activity (exo+) and AccuPrime Pfx DNA polymerase, which is used in this protocol, are suitable for PCR amplification involving Ds and Px. The $3′ → 5′$ exonuclease activity removes the unnatural substrates misincorporated opposite the natural

bases during PCR, which is important for the high Ds–Px pairing selectivity [8]. In the one-way incorporation of modified dPxTP opposite Ds in templates, TITANIUM Taq DNA polymerase can also be used [19].

15. Store the unnatural base substrate solution (dDsTP [6], dDiol1-PxTP [8], and dPa'TP [6, 12]) at –10 to –30 °C, and minimize repeated freeze–thaw cycles. Without repeated freeze–thaw cycles, these substrates are stable for at least a year. We generally dilute the original stock solution to a 0.5 or 0.25 mM working solution, and store aliquots with appropriate volumes.

16. To easily crush gel slices, use a disposable BioMasher stir bar (BioMasher II for a 1.5–2-ml tube and BioMasher V for a 50-ml conical tube), instead of a disposable pipette tip or a spatula.

17. The total PCR reaction volume depends on the amount of PCR products that will be sufficient for further use. In general, eight PCR (0.1-ml scale) reactions with a sufficient PCR cycle number will yield about 100–300 pmol of Ds-containing DNA library after purification on a denaturing gel. The amplification yield depends on both the number of PCR cycles and the amount of input DNA. Before large scale PCR amplification, assessing the appropriate number of PCR cycles is highly recommended by real-time PCR with a 25-μl reaction volume, in the presence of 30,000–75,000-fold diluted SYBR Green I (Lonza), to determine the number of PCR cycles sufficient for amplification. Too many PCR cycles might increase the unnatural to natural base mutations, as well as the amounts of undesired longer and/or shorter by-products.

18. We generally make a 6-well slot gel with a 2 mm thickness, and about 400 μl of the sample solution can be loaded per well. We load as much of the sample solution per well as possible into the middle slots, and load a marker dye on both sides.

19. Two bands will be detected at the upper position for xylene cyanol. The upper band corresponds to the Px-strand with an extra oligo-T (T15) sequence via the C12 alkyl spacer, and the lower band corresponds to the Ds-strand, i.e., the Ds-containing DNA library of interest. Depending on the number of PCR cycles employed, unconsumed primers might be detected at the positions between the xylene cyanol and the bromophenol blue.

20. Lower amounts of the Ds-containing DNA library can be used, with an appropriate increase in the PCR cycle number.

References

1. Ellington AD, Szostak JW (1990) In vitro selection of RNA molecules that bind specific ligands. Nature 346:818–822

2. Tuerk C, Gold L (1990) Systematic evolution of ligands by exponential enrichment: RNA ligands to bacteriophage T4 DNA polymerase. Science 249:505–510

3. Kong HY, Byun J (2013) Nucleic Acid aptamers: new methods for selection, stabilization, and application in biomedical science. Biomol Ther (Seoul) 21:423–434

4. Gupta S, Hirota M, Waugh SM, Murakami I, Suzuki T, Muraguchi M, Shibamori M, Ishikawa Y, Jarvis TC, Carter JD, Zhang C, Gawande B, Vrkljan M, Janjic N, Schneider DJ (2014) Chemically modified DNA aptamers bind interleukin-6 with high affinity and inhibit signaling by blocking its interaction with interleukin-6 receptor. J Biol Chem 289:8706–8719

5. Gold L, Ayers D, Bertino J, Bock C, Bock A, Brody EN, Carter J, Dalby AB, Eaton BE, Fitzwater T, Flather D, Forbes A, Foreman T, Fowler C, Gawande B, Goss M, Gunn M, Gupta S, Halladay D, Heil J, Heilig J, Hicke B, Husar G, Janjic N, Jarvis T, Jennings S, Katilius E, Keeney TR, Kim N, Koch TH, Kraemer S, Kroiss L, Le N, Levine D, Lindsey W, Lollo B, Mayfield W, Mehan M, Mehler R, Nelson SK, Nelson M, Nieuwlandt D, Nikrad M, Ochsner U, Ostroff RM, Otis M, Parker T, Pietrasiewicz S, Resnicow DI, Rohloff J, Sanders G, Sattin S, Schneider D, Singer B, Stanton M, Sterkel A, Stewart A, Stratford S, Vaught JD, Vrkljan M, Walker JJ, Watrobka M, Waugh S, Weiss A, Wilcox SK, Wolfson A, Wolk SK, Zhang C, Zichi D (2010) Aptamer-based multiplexed proteomic technology for biomarker discovery. PLoS One 5:e15004

6. Hirao I, Kimoto M, Mitsui T, Fujiwara T, Kawai R, Sato A, Harada Y, Yokoyama S (2006) An unnatural hydrophobic base pair system: site-specific incorporation of nucleotide analogs into DNA and RNA. Nat Methods 3:729–735

7. Kimoto M, Kawai R, Mitsui T, Yokoyama S, Hirao I (2009) An unnatural base pair system for efficient PCR amplification and functionalization of DNA molecules. Nucleic Acids Res 37:e14

8. Yamashige R, Kimoto M, Takezawa Y, Sato A, Mitsui T, Yokoyama S, Hirao I (2012) Highly specific unnatural base pair systems as a third base pair for PCR amplification. Nucleic Acids Res 40:2793–2806

9. Yang Z, Chen F, Alvarado JB, Benner SA (2011) Amplification, mutation, and sequencing of a six-letter synthetic genetic system. J Am Chem Soc 133:15105–15112

10. Malyshev DA, Dhami K, Quach HT, Lavergne T, Ordoukhanian P, Torkamani A, Romesberg FE (2012) Efficient and sequence-independent replication of DNA containing a third base pair establishes a functional six-letter genetic alphabet. Proc Natl Acad Sci U S A 109: 12005–12010

11. Kimoto M, Yamashige R, Matsunaga K, Yokoyama S, Hirao I (2013) Generation of high-affinity DNA aptamers using an expanded genetic alphabet. Nat Biotechnol 31:453–457

12. Mitsui T, Kimoto M, Sato A, Yokoyama S, Hirao I (2003) An unnatural hydrophobic base, 4-propynylpyrrole-2-carbaldehyde, as an efficient pairing partner of 9-methylimidazo[(4,5)-b]pyridine. Bioorg Med Chem Lett 13:4515–4518

13. Bock LC, Griffin LC, Latham JA, Vermaas EH, Toole JJ (1992) Selection of single-stranded-DNA molecules that bind and inhibit human thrombin. Nature 355:564–566

14. Mayer G, Hover T (2009) In vitro selection of ssDNA aptamers using biotinylated target proteins. Methods Mol Biol 535:19–32

15. Mosing RK, Bowser MT (2009) Isolating aptamers using capillary electrophoresis-SELEX (CE-SELEX). Methods Mol Biol 535:33–43

16. Navani NK, Mok WK, Yingfu L (2009) In vitro selection of protein-binding DNA aptamers as ligands for biosensing applications. Methods Mol Biol 504:399–415

17. Chai C, Xie Z, Grotewold E (2011) SELEX (Systematic Evolution of Ligands by EXponential Enrichment), as a powerful tool for deciphering the protein-DNA interaction space. Methods Mol Biol 754:249–258

18. Ogawa N, Biggin MD (2012) High-throughput SELEX determination of DNA sequences bound by transcription factors in vitro. Methods Mol Biol 786:51–63

19. Yamashige R, Kimoto M, Mitsui T, Yokoyama S, Hirao I (2011) Monitoring the site-specific incorporation of dual fluorophore-quencher base analogues for target DNA detection by an unnatural base pair system. Org Biomol Chem 9:7504–7509

Chapter 5

Capillary Electrophoresis for the Selection of DNA Aptamers Recognizing Activated Protein C

Nasim Shahidi Hamedani and Jens Müller

Abstract

Capillary electrophoresis-based SELEX (CE-SELEX) is an efficient technique for the isolation of aptamers binding to a wide range of target molecules. CE-SELEX has a number of advantages over conventional SELEX procedures such as the selection of aptamers can be performed on non-immobilized targets, usually within a fewer number of selection cycles. Here we describe a complete procedure of CE-SELEX using activated protein C (APC) as the target protein.

Key words CE-SELEX, Aptamer, Capillary electrophoresis, SELEX, Activated protein C

1 Introduction

Aptamers are single stranded DNA or RNA molecules which are able to bind to different target molecules ranging from small organic molecules to entire organisms. Aptamers are typically selected from randomized libraries of nucleic acids using a procedure termed *S*ystematic *E*volution of *L*igands by *Ex*ponential Enrichment (SELEX) which was introduced for the first time in 1990 [1, 2]. The SELEX-procedure consists of multiple rounds of selection, partitioning, and amplification which are repeated to allow for the enrichment of aptamers with high binding affinity. This procedure will be completed by cloning and/or sequencing, and evaluation of individual aptamer sequences [3].

During conventional SELEX, targets need to be immobilized onto solid supports to allow for efficient separation from non-binding ssDNA-molecules. However, further progressions led to the development of homogenous methods, such as capillary electrophoresis (CE)-SELEX, which allow the selection of aptamers against free targets [4].

Günter Mayer (ed.), *Nucleic Acid Aptamers: Selection, Characterization, and Application*, Methods in Molecular Biology, vol. 1380, DOI 10.1007/978-1-4939-3197-2_5, © Springer Science+Business Media New York 2016

In CE-SELEX, the random library is incubated with the target molecules in free solution and then the mixture containing free target molecules, target-ssDNA complexes and free ssDNA is injected into a capillary column and separated under high voltage. Collecting the outlet fraction at the retention time specific for target-bond ssDNA brings about the opportunity of gathering target-binding DNA-aptamers.

Besides homogeneous conditions, this kind of selection has additional advantages such as a high resolving power that reduces the number of cycles needed for selection to 4–6 cycles instead of 8–12 cycles when using conventional selection schemes [5]. However, also potential disadvantages do accompany CE-SELEX such as limitation in the total number of ssDNA-molecules introduced to the capillary or the difficulty of selecting aptamers against basic or low molecular weight target molecules [6].

In this chapter, a protocol for CE-SELEX of DNA-aptamers against activated Protein C (APC) is described. Although elucidated for the use of a ProteomeLab PA 800 System (Beckman Coulter, Krefeld, Germany), the described principles are also applicable when using other CE-systems.

2 Materials

2.1 Capillary Electrophoresis

1. Proteomelab™ PA 800 capillary electrophoresis (Beckman Coulter, Inc., Fullerton, CA, USA) equipped with UV/PDA detector.

2. Bare fused-silica capillary, 67 cm total length, 50 cm effective length, 50 μm inner diameter (i.D.), 375 μm outer diameter (o.D.) (Beckman Coulter, Inc. Brea, CA, USA).

3. Plastic vials, 0.5 ml.

4. Glass vials, 2 ml and caps (Beckman Coulter, Inc. Brea, CA, USA).

5. Random ssDNA-library IHT1: 5'-AAG CAG TGG TAA GTA GGT TGA-N_{40}-TCT CTT CGA GCA ATC CAC AC-3'. Order 1 μmol synthesis scale followed by PAGE purification. Store lyophilized powder at 2–8 °C until dissolved. Aliquot and store resolved stock solutions (e.g., 100 μM) at ≤–20 °C until used.

6. Separation buffer: 25 mM Tris–HCl, 10 mM NaCl, 1 mM KCl, 1 mM $CaCl_2$, and 1 mM $MgCl_2$, pH 8.3 (*see* **Notes 1** and **2**).

7. Human activated protein C (APC) (e.g., Haematologic Technologies, Essex Junction, Vermont, USA). Store stock solutions as indicated on label until used (*see* **Note 3**).

8. Vivaspin®6 centrifugal concentrators with 10,000 Da MW cut-off (Sartorius Stedim, Goettingen, Germany).

9. Washing buffers: 0.1 N NaOH; 0.1 N HCl; ultrapure water.

2.2 Polymerase Chain Reaction (PCR)

1. Thermal cycler.

2. HotStarTaq *Plus* DNA polymerase including buffers.

3. Amplification primers targeting the fixed sequences of the library in full length, HPLC purified. Store lyophilized powder at 2–8 °C until dissolved. Aliquot and store resolved stock solutions (e.g., 100 µM) at <–20 °C until used.

4. Deoxynucleotide triphosphates solution, 25 mM of each. Aliquot and store stock solutions at –20 °C until used.

5. PCR tubes 0.2 ml.

2.3 Agarose Gel

1. LE Agarose.

2. Tris Borate-EDTA buffer: 50 mM Tris, 45 mM boric acid, and 0.5 mM EDTA, pH 8.4.

3. Ten mg/ml ethidium bromide. Aliquot and store stock solutions. Add adequate amount of ethidium bromide to agarose cooled to 50–60 °C to reach a final concentration of 0.5 µg/ml (*see* **Note 4**).

4. DNA molecular weight marker XIII, 50 base pair ladder.

5. Loading buffer for gel electrophoresis: 40 % sucrose, 0.1 % xylene cyanol, and 0.1 % bromophenol blue. Store stock solutions at 4–8 °C until used.

2.4 ssDNA Production

1. NanoDrop® ND-1000 UV/Vis-Spectrophotometer (Thermo Scientific).

2. Thermomixer.

3. Magnetic beads separator.

4. Streptavidin-coated magnetic beads (SMB), Dynabeads M-280 Streptavidin (Life Technologies, Karlsruhe, Germany). Store the vial upright to keep the beads in liquid suspension since drying of the beads will result in reduced performance. Store the vial at 2–8 °C, avoid freezing.

5. 5′-biotinylated capture molecules, complementary to a part of the 3′ primer-binding section of the IHT1 library: 5′-Biotin TGG ATT GC-3′. Store lyophilized powder at 2–8 °C until dissolved. Aliquot and store resolved stock solutions (e.g., 100 µM) at –20 °C until used.

6. Binding and washing buffer 1 (B&W 1): 5 mM Tris–HCl, 1 M NaCl, 0.5 mM EDTA, pH 7.5.

7. Binding and washing buffer 2 (B&W 2): 5 mM Tris–HCl, 1 M NaCl, pH 7.5.

8. Washing buffer: 10 mM Tris–HCl, 20 mM NaCl, pH 7.5.

9. 5 M NaCl-solution.

10. Protease-free bovine serum albumin (BSA). Store at 4 °C.

11. SMBs storage buffer: 1× PBS, 0.1 % BSA, 0.02 % NaN_3, pH 7.4.

2.5 Filter Retention Analysis

1. Phosphorimager.

2. Dot-Blot system, e.g., Minifold® I Blotting System (Whatman, USA).

3. T4 polynucleotide kinase.

4. Phosphorimager screen and matching cassette.

5. γ-^{32}P ATP (PerkinElmer, Rodgau, Germany).

6. Dulbecco's phosphate buffered saline containing 0.5 mM $MgCl_2$, 0.9 mM $CaCl_2$.

7. Illustra microspin G-25 columns.

8. Nitrocellulose membranes, 0.45 μm pore size.

9. Yeast tRNA, 10 mg/ml.

3 Methods

During the first step of CE-SELEX, the randomized library is incubated with the target molecule (e.g., APC). After incubation, a small volume of the sample is injected into a primed, silica-fused capillary for CE-based separation of non-binding from target-bound sequences. The loaded capillary is then placed to span two reaction tubes filled with neutral to basic conductive buffer solution. During separation under high voltage applied between the tubes, positively charged buffer ions that are attracted to the negatively charged surface of the capillary do migrate to the cathodic end, resulting in a constant bulk flow of electrolytes that is called the electroosmotic flow (EOF) and represents the main trigger responsible for the mobility of injected materials within the capillary. Because the force of the EOF is greater than the electrophoretic mobility of the compounds, all injected molecules migrate from the inlet (anodic) to the outlet (cathodic) of the capillary [7]. Due to the electrophoretic attraction, the positively charged molecules move faster while negatively charged molecules are retained longer because of their contradictory electrophoretic mobilities. Therefore, depending on its mass and charge, each specific molecule possesses a specific retention time under the conditions defined by the electrophoresis setup [8]. The negatively charged ssDNA sequences which show binding affinity to the faster moving target protein molecules migrate at retention times that are shorter than that of the bulk non-binding ssDNA-library molecules. Thus, target-binding sequences can be collected from the outlet of the capillary within the so-called collection window that is the time between the start of the separation and the time that unbound sequences reach the outlet.

Collected sequences are amplified and generated single-strands introduced to the next round of the above described CE-SELEX. Usually 4–6 cycles of selection are required for the enrichment of an aptamer-pool showing peak bulk binding affinity. Subsequent analysis of included single aptamer-sequences by either cloning/Sanger-sequencing or next-generation sequencing approaches finally leads to the definition of candidate sequences to be tested for binding affinity by filter retention analysis.

Due to the lower amount of ssDNA that is injected into the capillary, in comparison to other selection methods, an increased risk of contamination with non-target-specific sequences stemming from capillary and/or instrument contaminations must be considered. The most critical source of contamination is the unbound library-sequences which migrate in spatial proximity to the desired specific aptameric sequences. As the amount of the specific sequences is trivial when compared to the bulk library sequences, contamination of the outlet of the capillary with non-binding sequences obviously reduces selection efficiency. Another source of contamination are the PCR-products from previous rounds of selection. Thus, rigorous separation of pre- and post-PCR areas as well as pre- and post-PCR materials is needed to avoid potential contaminations of evolved pools with previous-generation sequences.

Within the following sections, the main general procedures for the selection of DNA-aptamers against APC by CE-SELEX are described. Please consult the manual of the used CE-system/software for specific technical details.

3.1 Installation and Conditioning of a New Capillary

Install a new capillary for each individual selection. The following points describe the most critical steps during installation and use of a new capillary when running the Beckman Coulter PA 800 System.

1. Remove seal retainer clips as well as the aperture plug and the O-ring form the cartridge. Firmly remove the used capillary by pulling it out from the cartridge inlet side. Insert the new capillary into the outlet side of the cartridge with the end utmost from the capillary detection window (near to the cartridge window).

2. Push the capillary carefully into the cartridge base until it appears at the inlet. Protect detection window of the capillary from breakage (*see* **Note 5**).

3. Once the end of the capillary appears in the inlet side of the cartridge, pull it from the inlet side until the detection window appears centered within the cartridge window.

4. Insert the capillary seal clips over the capillary at both inlet and outlet side. Use the Capillary Length Template to accordingly cut both ends of the capillary using the cleavage stone. In doing

so, adjust the ends of the capillary to be 1 mm shorter than the electrodes within the final CE-Cartridge assembly. Then reinstall the aperture plug and O-ring.

5. Check the capillary ends under magnification and recut/readjust the capillary in case of angled or denticulated ends.

6. Condition the capillary before the first use. For silica-fused capillaries, use the conditioning program described in Table 1.

3.2 CE-Based Isolation of Target-Binding ssDNA-Molecules

1. Dilute the starting library in separation buffer to yield a concentration of 25 μM in final volume of 20 μl (*see* **Note 6**). Use a final concentration of 0.5 μM of selected ssDNA pools during the subsequent cycles (*see* **Note 7**).

3.2.1 Incubation of ssDNA-Library and APC

2. Heat the thus diluted library to 90 °C for 5 min using one single PCR tube and let it to return to the room temperature to allow for proper folding of the random ssDNA-molecules.

3. Centrifuge the PCR tube shortly.

4. Spike the APC target-protein into the ssDNA pool to reach the final concentration of 0.5 μM for the first cycle and incubate the mixture for 30 min at RT (*see* **Notes 8** and **9**).

3.2.2 Injection into Capillary and Separation of Components under EOF

1. Wash both ends of capillary and electrodes with distilled water and dry it using cotton swabs.

2. Place the single PCR tube containing target-ssDNA-mixture in injection site and prepare assembly needed for sample injection.

3. Add 100 μl of separation buffer each into tubes that will be defined and used as the inlet and outlet buffer vials during separation.

4. Adjust the separation temperature for the capillary to 20 °C.

5. Perform separation using a program as described in Table 2 (*see* **Note 10**).

Table 1
Conditions employed for capillary conditioning

Reagent	Pressure (psi)	Voltage	Duration (min)
NaOH, 0.1 M	20	–	4
Air drying	20	–	2
ddH$_2$O	20	–	2
Separation buffer	20	–	4
Separation buffer	–	15 kV with 2 min ramping time	6

Table 2
Conditions employed for CE separation

Step	Reagent	Pressure	Voltage	Duration	Mode of action
1	Target protein-ssDNA mixture	4 psi—inlet	–	5 s	Hydrodynamic Injection (*see* **Notes 11** and **12**)
2	Moving the inlet of capillary from injection vial to an inlet vial containing separation buffer				
3	Separation buffer	20 psi—both inlet and outlet	25 kV[a]	20 min[b]	Separation with the positive electrode at the inlet
4	End				

[a]Performing constant voltage should supply constant current during separation (*see* **Note 13**)
[b]Duration of separation depends on the retention time of the unbound fraction of ssDNA. Separation must be stop before the unbound ssDNA start to migrate out of the capillary. Determine collection window before start of the actual process for CE-SELEX (*see* **Note 14**)

6. Remove the collected fraction vial with caution using a new pair of gloves and close cap immediately as any contamination with non-binders will reduce the efficiency of selection.

3.2.3 Washing Process between the Runs

As the separation procedure stops before the migration of unbound sequences out of the capillary, a precise washing step is required to remove the unbound sequences from the capillary while protecting the instrument as well as the surrounding area from contamination by unspecific sequences (*see* **Note 15**).

1. Use the 0.5 ml plastic vials and buffer trays for washing step (*see* **Note 16**).

2. Clean the blue vial caps with distilled water and with aid of syringe.

3. Try to fill the vials starting from the bottom to avoid air bubbles.

4. All vials must be caped before starting the electrophoresis.

5. Follow the program indicated in Table 3.

3.3 PCR-Based Amplification of Selected ssDNA

1. Prepare a PCR master mixture containing 0.8 mM dNTPs, 1 mM each forward and reverse primer, 1.5 mM $MgCl_2$, 1.25 U/reaction HotStart*Taq* DNA polymerase and 20 µl of sample in a total volume of 100 µl.

2. Amplify collected ssDNA in a total of five reactions at 95 °C for initial activation of HotStart*Taq* DNA polymerase followed by 30 cycles of 95 °C for 30 s, 56 °C for 30 s, and 72 °C for 30 s.

3. Pool all PCR mixtures and check the quality by running 10 µl of PCR product mixed with 2 µl of 5× loading buffer on a 2 % agarose gel.

Table 3
Conditions employed for in between selection cycle washes

Reagent	Pressure (psi)	Voltage	Duration (min)	Mode of action
HCl, 0.1 M	20	–	5	Reverse rinse wash
NaOH, 0.1 M	20	–	5	Reverse rinse wash
ddH$_2$O	20	–	5	Reverse rinse wash
Washing buffer	20	–	5	Reverse rinse wash

3.4 Asymmetric PCR and Isolation of ssDNA

The production of ssDNA is a crucial step of the SELEX-process. This paragraph describes the application of 'Capture and Release' (CaR) for the isolation of ssDNA from asymmetric PCR mixture (*see* **Note 17**) [9]. During the approach described here, additional asymmetric PCR is performed on previously amplified selected ssDNA (*see* Subheading 3.2).

1. Dilute the yielded PCR products (*see* Subheading 3.2) 1 in 10 using distilled water.

2. Add 10 µl of the dilution to 10 PCR vials containing 90 µl of asymmetric PCR master mixture (prepared as described in Subheading 3.2 but without addition of reverse primers).

3. Perform reactions in a thermal cycler by applying 50 cycles of the temperature profile described in Subheading 3.2. Check the quality of ssDNA obtained from asymmetric amplification by running a 10 µl sample on a 2 % agarose gel.

4. Resuspend the streptavidin magnetic beads by shaking the vial vigorously and take 1 mg of the beads (100 µl of 10 mg/ml stock suspension).

5. Wash the beads three times using B&W 1 buffer and a suitable magnetic device. Incubate the beads in 200 µl B&W 1 containing 1 µM of capture molecules (2 µl of 100 µM stock solution) for 30 min at room temperature. Prevent settling of the beads by shaking vigorously during incubation.

6. Wash the beads three times using 1 ml of B&W 2.

7. Pool and add the total of 1 ml of the asymmetric reaction mixtures to the loaded SMBs followed by spiking with 5 M NaCl to reach a final concentration of 100 mM.

8. Incubate for 30 min at room temperature. Prevent settling of the beads by shaking vigorously during incubation.

9. Wash the beads three times using washing buffer.

10. Add 20 μl of preheated purified water to the beads and incubate for 2 min at 43 °C to release captured ssDNA. Collect supernatant after separation of beads at 43 °C.

11. Determine the concentration of obtained ssDNA by NanoDrop UV-measurement.

12. Use isolated ssDNA for the next selection cycle.

13. For storage, resuspend the loaded SMBs in storage buffer and store at 4 °C until used.

3.5 Filter Retention Experiment

1. Add 5–10 pmol of purified ssDNA to a master mixture containing 5 μl T4 PNK buffer (10×), 2 μl T4 polynucleotide kinase (T4 PNK, 10 U/μl), 2 μl Gamma ^{32}P ATP (3.3 μM; 10 μCi/μl) in a final volume of 50 μl.

2. Incubate the mixture for 30 min at 37 °C.

3. Prepare the G-25 columns by resuspending the resin by vortexing. Twist off the bottom closure and centrifuge at $735 \times g$ for 1 min.

4. Pipet the labeling reaction to the top-center of the resin. Avoid disturbing the resin bed (*see* **Note 18**).

5. Purify the labeling reaction by centrifugation at $735 \times g$ for 2 min. Discard used G-25 column.

6. Check the removal of unbound radioactivity as well as the integrity of the labeled DNA by PAGE-analysis.

7. Dilute the ^{32}P-labeled DNA 1:10 with 1× D-PBS, heat it up to 90 °C for 5 min followed by cooling down to room temperature (*see* **Note 19**). This temperature treatment is necessary for obtaining stable conformation of ssDNA at room temperature.

8. For each aptamer pool or single sequence to be tested, prepare a dilution series of the target protein (APC) in D-PBS containing 0.1 % BSA and 10 μM yeast t-RNA. Pipette 24 μl of each dilution into a single well of a microtiter-plate. Always include a buffer-only sample. Run all analysis in at least duplicated. Add 1 μl of pre-diluted ^{32}P-labeled to each of the designated wells.

9. Cover the plate using Parafilm and incubate it at 37 °C for 30 min.

10. Soak the nitrocellulose membrane in freshly prepared 0.4 M NaOH followed by washing with 1× D-PBS (without BSA and tRNA).

11. Transfer the pretreated nitrocellulose membrane into the Dot-Blot system, apply the vacuum and wash each well three times using 150 μl of 1× D-PBS.

12. Use an 8-channel pipette to transfer 20 μl of the incubation mixtures to individual wells of the prepared blotting assembly.

13. Wash each well three times using 150 μl 1× D-PBS to remove non-target-bound sequences.

14. Remove membrane from the device and allow to air dry.

15. Pipett 0.8 μl of the used dilution of each applied ^{32}P-labeled DNA onto the same or another nitrocellulose membrane. These spots represent the total amount of radioactivity (i.e., labeled DNA) that was introduced to each well. Cover membranes by using a thin plastic foil, assemble with screen and close cassette (*see* **Note 20**).

16. Scan screen using the phosphorimager and quantify the single dots relative to the corresponding 100 % spots (Fig. 1).

17. Use 4-parametric regression analysis for calculation of K_D-values. Sigmoidal curve patterns are needed to yield reliable results.

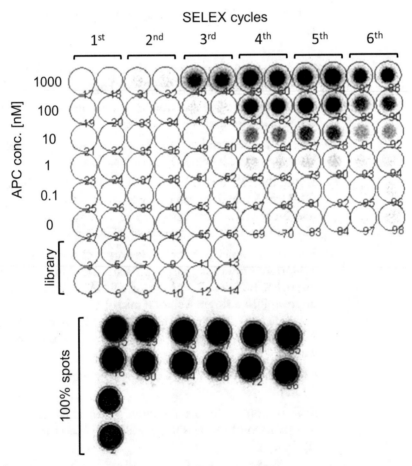

Fig. 1 Example of Dot-Blot images for the assessment of pool binding affinities over performed selection cycles. For testing of the starting random library, the indicated APC-concentrations were applied in a transposed pattern (wells 3–14). The 100 % spots were arranged in accordance with the according to the applied sample wells. The shown data revealed highest percentage of binding and affinity of the aptamer pool gathered after five cycles of selection

4 Notes

1. All buffers should be prepared in ultrapure water using analytical grade reagents. Buffers should be filtered before use as present particles may interfere with proper electroosmotic flow or even plug the used capillary.

2. The composition of the buffer and the salt concentration can be varied according to the selection conditions. The current set by a certain voltage is a function of salt concentration in the separation buffer. However, exceeding more than the maximum tolerated current (300 mA) can harm the interface block of the instrument.

3. As changing of the composition of the used selection buffer can not only interfere with aptamer-binding but also produce fluctuations in the electropherogram, we strongly recommend to change the buffer of the used APC using Vivaspin®6 concentrators by three consecutive additions of the separation buffer in the same volume as introduced APC followed by centrifugation at $2000 \times g$ at 4 °C. The resulting APC concentration might be determined using NanoDrop® ND-1000 UV/Vis-Spectrophotometer regarding to the APC extinction coefficient of $E^{1\%}_{1cm,280nm} = 14.5$.

4. The major drawback of ethidium bromide is its mutagenic potential. When used, ethidium bromide solution should therefore be handled with caution.

5. Always use gloves while installing a new capillary as finger print reduces detection sensitivity in capillary window. Handle the capillary with caution as the detection window is fragile.

6. No peak in electropherogram might be a sign of an air bubble at the bottom of the sample vial or insufficient quantity of sample in sample microvial. In these cases, remove the air bubble or increase the sample volume up to 100 μl, respectively.

7. The concentration of random library has a significant impact on the initial number of unique sequences introduced to the target and the capillary. Due to the presence of multiple copies of binding sequences after the first selection cycle, a reduced concentration of ssDNA (e.g., 0.5 μM) during the consecutive selection cycles will not affect the efficiency of the selection.

8. Injection of a sample with a different salt concentration than that of the separation buffer produces fluctuation in baseline of the electropherogram. Therefore, ensure that the incubation mixture of the ssDNA and the target protein corresponds to the composition of the selection buffer.

9. Gradually reduce the protein concentration with each selection cycle. It has been previously shown that the efficiency of enrichment of best binders is directed by the stringency of the selection that is increased by reduction of the protein concentration [10]. As the target concentration decreases, the presence of high affinity aptamers in the collected pool increases while the considerable point is the practical lower limit which means that once the target protein concentration drops below the K_D value of the aptamer with highest binding affinity, further decreasing in the protein concentration has no significant or only a slight impact on further enrichment [11].

10. Under a certain condition applied to the capillary and by using the same buffer system, each target protein, bound fraction of the library-target protein and unbound ssDNA migrate at a certain time. In case of APC as the target protein, determination of the collection window required separate injections of protein and library in order to assess individual retention times (Fig. 2). In our opinion, collection of target-binding sequences should be already stopped when the bulk library sequences become detectable by UV-measurements.

11. There are two different possibilities for samples injection: (1) Hydrodynamic injection (2) Electrokinetic injection. In hydrodynamic injection, an applied pressure for a certain time introduces the sample to the capillary column which is known as the most frequently used injection technique. In electrokinetic injection, an applied current or voltage for a certain time causes the sample to migrate into the capillary column. This kind of injection is mostly applied for high viscosity materials which is not common in CE-SELEX [12].

12. The volume of the sample introduced to the capillary (V_{inj}) by hydrodynamic injection is a function of the capillary inner diameter, the viscosity of the buffer, the applied pressure and injection time. The loaded volume can be calculated using the Hagen–Poiseuille equation [8]:

$$V_{inj} = \frac{P d^4 \pi t_{inj}}{128 \eta L}$$

Where
ΔP = pressure difference across the capillary
d = capillary inside diameter
t_{inj} = injection time
η = buffer viscosity
L = total capillary length.

13. Low or unsteady current might be an indicator of a plugged capillary. One solution is to rinse the capillary with ddH$_2$O at

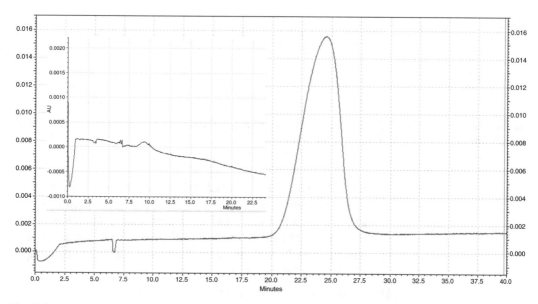

Fig. 2 Determination of aptamer collection window. Electropherograms obtained from separate injections (4 psi, 5 s) of the IHT1 random library (25 μM) with UV detection at 254 nm and activated protein C (2 μM) with UV detection at 214 nm (*inset*)

100 psi for 10 min. Change the capillary to a new one in case the problem persists [13].

14. As the retention time indicated in an electropherogram is the migration time of the define compound to the detection window (*see* Fig. 3) but not to the end of the capillary, the exact time at which each compound reaches to the end of the capillary can be calculated using equation below:

$$x = t_m + (\frac{l_E}{l_D} t_m)$$

l_D: Capillary length to the detector or effective length
l_E: Capillary length from detection window to the end (which for Beckman Coulter PA 800 capillaries is constant to 10.2 cm)
t_m: Migration time of the defined compound to the detection window.

15. Contamination of collected fractions with unspecific sequences should be considered in SELEX procedures using capillary electrophoresis. Due to the small amount of molecules injected into the capillary, contamination of the outlet with the bulk unbound sequences can negatively interfere with the next

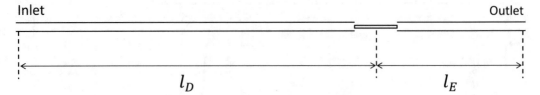

Inlet   Outlet

$$l_D \qquad\qquad\qquad l_E$$

Fig. 3 Schematic presentation of the capillary window position. l_D: Capillary length to the detection window, l_E: Capillary length from detection window to the outlet

SELEX cycle. To avoid such a huge source of contamination, a strategy that prevents the bulk library sequences from reaching the outlet of the capillary is a necessity. Therefore, stopping the EOF during retention of the unbound sequences and flashing out these unbound sequences via the inlet of the capillary is a reasonable measure.

16. As the inlet of the capillary always comes into contact with unbound ssDNA at high concentrations during injection, it should always be assumed as a source of contamination. Thus, physical separation of vials/rubbers used at the inlet or the outlet of the capillary is strongly recommended. Furthermore, the use of single-use plastic vials instead of glass vials will significantly reduce potential sources of contamination.

17. Capture and Release (CaR) is an efficient procedure for isolation of ssDNA required for each selection cycle. Briefly, short biotinylated oligodeoxynucleotides, that are complementary to the 3′-end of the target single stranded oligodeoxynucleotides produced during asymmetric PCR, are bound to streptavidin magnetic beads. Incubation of the loaded streptavidin magnetic beads with asymmetric PCR mixtures results in capturing of the ssDNA which can be easily released after washing by altering temperature and ion strength conditions [9].

18. Although the G-25 columns may be applied in conjunction with a fixed-angle rotor, we observed much better performance when using a swing-out rotor.

19. The dilution factor depends on the intensity of radioactivity. For example, after one half life of applied or available radioactivity (14.3 days), the dilution factor may be reduced to 1:5.

20. The incubation time depends on the intensity of the radioactivity that retained on the nitrocellulose membrane. For samples freshly labeled with only little decayed ^{32}P, incubation for a few hours might be sufficient while low levels of radioactivity may necessitate an overnight incubation.

References

1. Tuerk C, Gold L (1990) Systematic evolution of ligands by exponential enrichment: RNA ligands to bacteriophage T4 DNA polymerase. Science 249:505–510

2. Ellington A, Szostak J (1990) *In vitro* selection of RNA molecules that bind specific ligands. Nature 346:818–822

3. Berezovski M, Drabovich A, Krylova SM, Musheev M, Okhonin V, Petrov A et al (2005) Nonequilibrium capillary electrophoresis of equilibrium mixtures: a universal tool for development of aptamers. J Am Chem Soc 127(9):3165–3171

4. Mendonsa SD, Bowser MT (2004) In vitro selection of high-affinity DNA ligands for human IgE using capillary electrophoresis. Anal Chem 76(18):5387–5392

5. Mosing RK, Mendonsa SD, Bowser MT (2005) Capillary electrophoresis-SELEX selection of aptamers with affinity for HIV-1 reverse transcriptase. Anal Chem 77(19):6107–6112

6. Ashley J, Li FY (2013) Three-dimensional selection of leptin aptamers using capillary electrophoresis and implications for clone validation. Anal Biochem 434:146–152

7. Whatley H (2001) Basic principles and modes of capillary electrophoresis. In: Petersen J, Mohammad AA (eds) Clinical and forensic applications of capillary electrophoresis. Humana Press Inc., Totowa, NJ

8. Lauer HH, Rozing GP (2009) High performance capillary electrophoresis: a primer. Agilent Technologies Inc., Waldbronn

9. Hamedani NS, Blümke F, Tolle F, Rohrbach F, Rühl H, Oldenburg J et al (2015) Capture and Release (CaR): a simplified procedure for one-tube isolation and concentration of single-stranded DNA during SELEX. Chem Commun 51:1135–1138

10. Ravelet C, Grosset C, Peyrin E (2006) Liquid chromatography, electrochromatography and capillary electrophoresis applications of DNA and RNA aptamers. J Chromatogr A 1117:1–10

11. Wang J, Rudzinski JF, Gong Q, Soh HT, Atzberger PJ (2012) Influence of target concentration and background binding on *in vitro* selection of affinity reagents. PLoS One 7(8):e43940

12. Agilent 7100 capillary electrophoresis system - users manual, Agilent Technologies, Inc. 2009

13. ProteomeLab™ PA 800 user's information, Beckman Coulter, Inc., Copyright 2004.

Chapter 6

Preparation of SELEX Samples for Next-Generation Sequencing

Fabian Tolle and Günter Mayer

Abstract

Fuelled by massive whole genome sequencing projects such as the human genome project, enormous technological advancements and therefore tremendous price drops could be achieved, rendering next-generation sequencing very attractive for deep sequencing of SELEX libraries. Herein we describe the preparation of SELEX samples for Illumina sequencing, based on the already established whole genome sequencing workflow. We describe the addition of barcode sequences for multiplexing and the adapter ligation, avoiding associated pitfalls.

Key words Aptamer, SELEX, Next-generation sequencing, High-throughput sequencing, Illumina dye sequencing

1 Introduction

The human genome project and other whole genome sequencing projects have created the need for higher throughput than conventional Sanger sequencing could offer [1, 2]. This lead to the development of a new generation of sequencing technologies known as next-generation (NGS), high-troughput (HTS), or deep sequencing. The high adoption rates of such techniques have lead to a dramatic per base price drop over the last decade. This degree of affordability has made NGS very attractive to the SELEX community [3, 4].

Traditionally the strongly enriched library from the final round of an in vitro selection experiment is subjected to cloning and Sanger sequencing. Usually 30–100 clones are analyzed, ideally revealing families of similar sequences that can be tested for binding [5].

With NGS up to 100 million sequences can be sequenced in one experiment, enabling a much deeper insight into the "black

Günter Mayer (ed.), *Nucleic Acid Aptamers: Selection, Characterization, and Application*, Methods in Molecular Biology, vol. 1380, DOI 10.1007/978-1-4939-3197-2_6, © Springer Science+Business Media New York 2016

box" of in vitro selection. The ability to multiplex several samples in one sequencing run enables deep sequencing, not only of the last round, but of each round of the entire SELEX. The ability to track sequence families and their amplification behavior over several rounds of selection allows much better prediction of interesting sequences, thereby increasing the success chances of in vitro selection experiments [6, 7].

In this chapter we describe the sample preparation of SELEX samples for Illumina sequencing.

In the first step we describe the PCR reaction for the amplification and the addition of "barcode" sequences, which are required for multiplexing, to the samples. With a specific index sequence for every selection round all samples can be sequences in one run. Later the data can be de-multiplexed and assigned to the right selection round.

In the second step the addition of adapter sequences by enzymatic ligation is described. The adapter sequences are needed for the immobilization and processing of the sample on the Illumina instruments. The process described uses parts of a kit available from Illumina, originally developed for the preparation of whole genome sequencing samples. By adapting the sample preparation to this already established workflow, a fast and robust sample preparation is achieved, not only for laboratories already familiar with whole genome sequencing workflows. The addition of indexed adapters allows an additional level of multiplexing, if a very cost effective sequencing of many samples is desired.

2 Materials

Prepare all solutions using ultrapure water and analytical grade reagents. Prepare all reagents at room temperature (unless indicated otherwise). Store all reagents at −20 °C (unless indicated otherwise).

2.1 PCR

1. Proof reading Polymerase such as *Pfu* or *Pwo* and the appropriate buffer (*see* **Note 1**).

2. dNTPs (25 mM of each nucleotide).

3. DNA template from the start library and each round of SELEX.

4. Indexed forward primer: 5′-Index-Forward-Primer-Sequence-3′.

5. Indexed reverse primer: 5′-Index-Reverse-Primer-Sequence-3′.

The following index sequences are recommended, allowing for up to 12 different SELEX rounds to be sequenced on one lane (*see* **Note 2**):

Index 1 ATCACG,	Index 2 CGATGT,	Index 3 TTAGGC,	Index 4 TGACCA,
Index 5 ACAGTG,	Index 6 GCCAAT,	Index 7 CAGATC,	Index 8 ACTTGA,
Index 9 GATCAG,	Index 10 TAGCTT,	Index 11 GGCTAC,	Index 12 CTTGTA.

2.2 DNA Purification and Concentration

1. Silica-based spin column kit (*see* **Note 3**) (e.g., NucleoSpin Clean-Up kit from Macherey-Nagel, Germany).

2.3 Agarose Gel Electrophoresis

1. Agarose (e.g., Agarose LE, Genaxxon) and ethidium bromide (1 μg per ml agarose solution).

2. TBE buffer (10×): Dissolve 108 g Tris and 55 g boric acid in 900 ml water. Add 40 ml 0.5 M Na_2EDTA (pH 8.0) and adjust volume to 1 L. Store at room temperature.

3. DNA loading buffer (6×): 60 % glycerol, 10 mM Tris, 60 mM Na_2EDTA (pH 8.0), 0.03 % xylene cyanol. Store at room temperature.

4. DNA ladder (e.g., GeneRuler Ultra Low Range DNA Ladder, Thermo Scientific).

2.4 Adapter Ligation

1. TruSeq DNA PCR-Free (LT) sample preparation kit (Ref.15037063, Illumina) including resuspension buffer (10 mM Tris–HCl (pH 8.5) with 0.1 % Tween).

3 Methods

3.1 PCR for Sample Amplification and Introduction of Index

Carry out all PCR sample preparation procedures on ice.

1. Prepare a PCR master mix for 13 (12 samples and a "No Template Control" (NTC)) individual 200 μl PCR samples according to Table 1.
2. Label and fill 13 PCR tubes with 195 μl of the master mix.
3. Add 2 μl of forward-primer (50 μM), 2 μl reverse-primer (50 μM) and 1 μl DNA template (ca. 0.1 μM). Use 1 μl water for the NTC.

Make sure to use primer pairs with the same index sequence.

4. Run PCR with your regular program; depending on the template concentration ca. 7–9 PCR cycles should be sufficient. Take care not to "over-amplify" to avoid by-product formation.
5. Check the PCR products on an analytical 4 % agarose gel.

Table 1
Protocol for setting up the index-PCR

Reagent	Stock conc.	Volume	Final conc.
Water		2394 µl	
Pwo-Buffer	10×	280 µl	1×
dNTPs	25 mM	28 µl	250 µM
Pwo-Polymerase	2.5 U/µl	28 µl	2.5 U

3.2 Analytical Agarose Gel Electrophoresis

1. Prepare a 4 % agarose gel (*see* **Notes 4** and **5**).
2. Take 5 µl PCR product and add 1 µl 6× DNA loading buffer.
3. Load the samples on the 4 % agarose gel and let it run for 10 min at 130 V.

3.3 DNA Purification with Spin-Column Kit (e.g., for NucleoSpin Clean-up Kit)

1. Mix 200 µl PCR product with 400 µl binding buffer (2 volumes).
2. Load each sample (600 µl) on one column.
3. Centrifuge the samples (1 min; 12,000 rcf), thereby binding the DNA on the silica matrix.
4. Discard the flow-through; add 650 µl wash buffer and centrifuge (1 min; 12,000 rcf).
5. Place columns in a fresh vial and centrifuge again (1 min; 12,000 rcf) to dry the column from any excess of wash buffer.
6. Place columns in a fresh, correctly labeled, vial.
7. Add 60 µl resuspension buffer supplied with the TruSeq DNA kit.
8. Incubate the silica membrane for 1 min.
9. Centrifuge to elute the DNA (1 min; 12,000 rfc).

3.4 Sample Blending

1. Measure the concentration of each purified sample in a photometer.
2. Mix equimolar DNA amounts from up to 12 samples with different indices for a total of 1–2 µg DNA in 60 µl final volume.
3. If the concentration is not high enough, concentrate the sample with the spin column kit and elute in a final volume of 60 µl resuspension buffer.

3.5 Library Preparation

For the ligation of the adapters, with some adaptations, the TruSeq DNA PCR-Free Sample Preparation Kit LT can be used. However, keep in mind that the kit is intended for the preparation of samples from genomic DNA and therefore, for SELEX sample preparation only three out of the six procedures apply ("End Repair", "Adenylation" and "Adapter Ligation"). "Fragmentation", "Clean Up" and "Size Selection" are *not* used. Please refer to the kits documentation for additional information.

3.6   End Repair

1. Add 40 µl End Repair Mix 2 to 60 µl of the purified PCR sample (*see* **Note 6**).

2. Run the "ERP" protocol in a thermocycler (30 min 30 °C; Hold 4 °C).

3.7   Purification

1. Purify the sample with the silica based spin-column kit as described previously (*see* **Note 7**).

2. Elute the sample in 40 µl resuspension buffer.

3.8   Adenylation of 3′ End

1. Add 20 µl A-Tailing-Mix to the sample.

2. Run the "ATAIL70" protocol in a thermocycler (30 min 37 °C; 5 min 70 °C; 5 min 4 °C).

3.9   Adapter Ligation

The adapter sequences are 63–65 nt long, therefore the desired product with adapters on both ends should be around 125 bp longer than the original PCR-product (*see* Fig. 1).

1. Add 5 µl Ligation-Mix 2.

2. Add 5 µl of the desired adapter (if several samples are multiplexed, note the adapter number!).

3. Run the "LIG" protocol in a thermocycler (10 min 30 °C; 5 min 4 °C).

3.10   Preparative Gel Purification

1. Prepare a 2 % agarose gel with large loading pockets.

2. Add 10 µl 6× loading buffer to the sample.

3. Load the sample into two pockets of the agarose gel (40 µl each).

4. Run the gel for around 60 min at 100 V (*see* **Note 8**).

5. Cut out the highest band with a clean scalpel as indicated in Fig. 2 (*see* **Notes 9** and **10**).

3.11   Purification

1. Purify the sample with the silica based spin-column kit.

2. Elute the sample in 40 µl resuspension buffer.

3. Run an analytical 4 % agarose gel to verify the success of the preparative gel purification.

3.12   Library Validation

1. To achieve highest sequencing data quality an accurate quantification of the DNA is necessary. This is done by qPCR with the KAPA library quantification kit. The PCR will only amplify samples with correctly annealed adapters on both ends, thereby confirming the quality of the sample.

3.13   Sequencing Run

1. After validation and quantification of the samples, they are ready to be sequenced on HiSeq or MiSeq instruments.

2. 75 bp single end sequencing is usually sufficient. However, for very long libraries 100 bp runs are recommended.

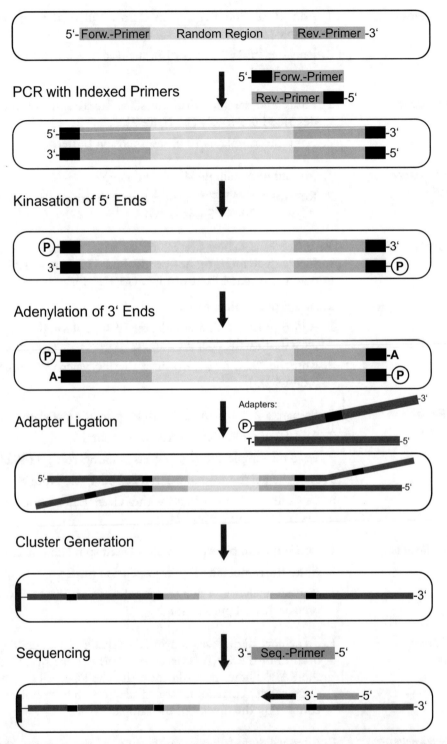

Fig. 1 Schematic overview of sample preparation and sequencing. Each SELEX sample is amplified with a different pair of special primers, containing a 6 nt long index region (depicted in *black*) at the 5′ end. For the adapter ligation, 5′-phosphates and 3′-deoxyadenosine-adenine overhangs are enzymatically added to the indexed PCR product. In the next step, adapters are ligated to both ends of the PCR product. The adapters contain DNA sequences necessary for the binding of the sample to the flow cell of the sequencer and the primer binding sites of the sequencing primers. Adapters may also contain indexsequences for an additional level of multiplexing

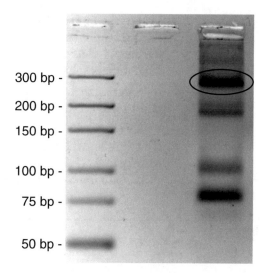

Fig. 2 Preparative agarose gel after adapter ligation. Usually the highest running band is the desired PCR product with adapters ligated to both ends (ca. 125 bp longer than the original PCR product). In most cases also undesired by-products are obtained, representing adapter dimers or PCR product without or with only one adapter. In case of doubt, cut out all bands and validate the correct product by qPCR

4   Notes

1. Any proofreading polymerase can be used. The addition of 3′-deoxyadenosine overhangs is not necessary at this step, therefore *Taq* polymerase is not recommended.

2. For technical reasons it is recommended to use as many different index sequences as possible, therefore we recommend the use of all 12 index sequences.

3. Spin column kits are available from many different suppliers. However please verify the suitability for agarose gel extraction in the handbook of your particular kit.

4. Use a magnetic steering bar to remove all bubbles to avoid excessive foaming of the gel in the microwave.

5. Gels can be stored for several weeks in 1× TBE at 4 °C.

6. T4 polynucleotide kinase from New England Biolabs (Frankfurt, Germany) can be used instead. However, better results were obtained with the "end repair mix" included in the kit.

7. *Do not* use the size selection beads provided with the kit, since they are optimized for a different size range and your desired product will not be collected.

8. Due to the buffer composition the samples may show an atypical running behavior. If the bands are not separated well after 60 min of electrophoresis, let the gel run longer until good separation is achieved.

9. If preparing several samples at the same time, take great care to avoid cross-contamination. Use a fresh scalpel for every sample.

10. If in doubt which band corresponds to the desired product with ligated adapters on both end, cut out all bands and check all samples for correct adapter ligation by qPCR.

Acknowledgements

The authors thank Dr. Marc Beyer, Dr. Michael Blank, Franziska Pfeiffer, and Silvana Hassel for the help in developing this protocol and critical reading of the manuscript.

References

1. Lander ES, Linton LM, Birren B et al (2001) Initial sequencing and analysis of the human genome. Nature 409:860–921

2. Abecasis GR, Auton A, Brooks LD et al (2012) An integrated map of genetic variation from 1,092 human genomes. Nature 491:56–65

3. Schütze T, Wilhelm B, Greiner N et al (2011) Probing the SELEX process with next-generation sequencing. PLoS One 6, e29604

4. Hoon S, Bin Z, Janda KD et al (2011) Aptamer selection by high-throughput sequencing and informatic analysis. Biotechniques 51:413–416

5. Stoltenburg R, Reinemann C, Strehlitz B (2007) SELEX—a (r)evolutionary method to generate high-affinity nucleic acid ligands. Biomol Eng 24:381–403

6. Cho M, Xiao Y, Nie J et al (2010) Quantitative selection of DNA aptamers through microfluidic selection and high-throughput sequencing. Proc Natl Acad Sci U S A 107:15373–15378

7. Hoinka J, Berezhnoy A, Dao P et al (2015) Large scale analysis of the mutational landscape in HT-SELEX improves aptamer discovery. Nucleic Acids Res 43(12):5699–5707

Chapter 7

Next-Generation Analysis of Deep Sequencing Data: Bringing Light into the Black Box of SELEX Experiments

Michael Blank

Abstract

In silico analysis of next-generation sequencing data (NGS; also termed deep sequencing) derived from in vitro selection experiments enables the analysis of the SELEX procedure (Systematic Evolution of Ligands by EXponential enrichment) in an unprecedented depth and improves the identification of aptamers. Besides quality control and optimization of starting libraries, advanced screening strategies for difficult targets or early identification of rare but high quality aptamers which are otherwise lost in the in vitro selection experiments become possible. The high information content of sequence data obtained from selection experiments is furthermore useful for subsequent lead optimization.

Key words Next generation sequencing, NGS, Bioinformatics, In silico analysis, SELEX

1 Introduction

NGS was originally developed for the purpose of whole genome sequencing and re-sequencing. At an early stage, scientists from other areas of basic, applied, and medical research realized the potential of the groundbreaking NGS technology and started to transfer it to their own research fields [1]. This trend can also be seen in the area of SELEX [2–12] and other DNA encoded ligand discovery technologies [13–16]. NGS and subsequent computational data analysis are applied to extract valuable information from starting and enriched combinatorial libraries, thereby opening new avenues in the field of aptamer identification and optimization. COMPAS (COMmonPAtternS) is a software tool that was developed to support the entire SELEX workflow at all relevant stages [2]. The COMPAS-analysis starts with the parsing process. Parsing prepares the raw NGS datasets (each comprising millions of sequence reads) for further analysis. In the next step, the parsed

Günter Mayer (ed.), *Nucleic Acid Aptamers: Selection, Characterization, and Application*, Methods in Molecular Biology, vol. 1380, DOI 10.1007/978-1-4939-3197-2_7, © Springer Science+Business Media New York 2016

data are analyzed in order to generate results that support and upgrade SELEX at different stages. A systematic method how NGS can be used to upgrade the SELEX procedure by subsequent COMPAS analysis is described in the following chapter.

2   Analysis of Deep Sequencing Data Derived from SELEX Experiments

The three main application fields are: (1) quality control of the starting libraries, (2) the improved identification of aptamers, and (3) aptamer optimization (Fig. 1a–c). In summary the contribution of in silico analysis to the in vitro selection procedure leverages SELEX and will finally enable access to aptamers having the desired properties.

2.1   Parsing of Raw NGS Data

The process that structures and organizes the millions of sequences derived from starting as well as from enriched SELEX rounds is termed parsing process. Raw sequences are checked for their usability to be further analyzed by various (bio)informatic and statistical algorithms. Exploitable sequences are filed in a way that individual SELEX libraries (starting as well as enriched libraries) are saved as individual data set files, which can be individually as well as comparatively analyzed. The focus is here on the variable data region that can be extracted as its beginning and end can be determined by the flanking, constant primer sites. Raw NGS data typically comprise so called FASTQ files. Besides sequence information, such files contain quality information to assess the accuracy of a sequencing run. They indicate the probability that a base was incorrectly determined at a given position by the sequencer. This Phred score value can be used as a quality measure to decide if an individual sequence is accepted or excluded for analysis. After the parsing process the individual starting and enriched SELEX libraries are filed into individual datasets each of them comprising millions of exploitable sequences.

2.2   Quality Control of Starting Libraries

Combinatorial SELEX starting libraries are assumed to provide a high number of different molecules forming a high variety of different three-dimensional structures that bind their target by complementary shape recognition. Typical starting libraries are synthesized with the goal of an equal distribution of nucleotide bases within in the random region. This means A, T, C, and G should be represented at 25 % at each position. However NGS and COMPAS analysis of a multitude of starting libraries that have been synthesized by different academic labs and commercial providers revealed that equally distributed SELEX libraries are rather a rare exception than the rule. In such biased starting libraries defined nucleotides are overrepresented at the cost of others. The distribution of nucleotides has presumably a direct influence on

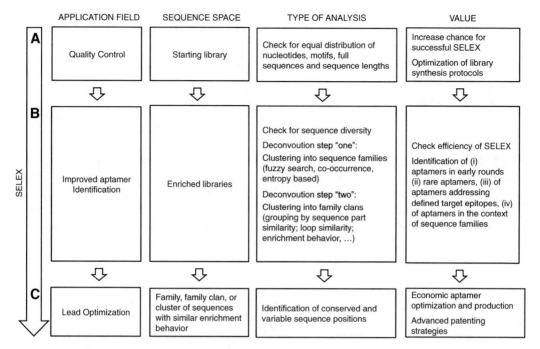

	APPLICATION FIELD	SEQUENCE SPACE	TYPE OF ANALYSIS	VALUE
A	Quality Control	Starting library	Check for equal distribution of nucleotides, motifs, full sequences and sequence lengths	Increase chance for successful SELEX Optimization of library synthesis protocols
B	Improved aptamer Identification	Enriched libraries	Check for sequence diversity Deconvoution step "one": Clustering into sequence families (fuzzy search, co-occurrence, entropy based) Deconvoution step "two": Clustering into family clans (grouping by sequence part similarity; loop similarity; enrichment behavior, ...)	Check efficiency of SELEX Identification of (i) aptamers in early rounds (ii) rare aptamers, (iii) of aptamers addressing defined target epitopes, (iv) of aptamers in the context of sequence families
C	Lead Optimization	Family, family clan, or cluster of sequences with similar enrichment behavior	Identification of conserved and variable sequence positions	Economic aptamer optimization and production Advanced patenting strategies

Fig. 1 Scheme giving an overview of COMPAS analysis of NGS data derived from SELEX experiments. COMPAS was designed to support the SELEX workflow at different stages: (**a**) quality control of starting libraries, (**b**) aptamer identification, and (**c**) aptamer optimization

the success rate of SELEX experiments as stretches of nucleotides build up motifs, which finally mediate binding to the respective target molecule: The more evenly the distribution of nucleotide building blocks, the more randomly is the distribution of motifs and the more diverse is the sequences space of the starting library. Consequently, the chance to end up with aptamer ligands in the SELEX process which have the expected binding characteristics increases with equally distributed starting libraries. The COMPAS software enables both, the analysis of the distribution of building blocks (Fig. 2a) as well as the analysis of distribution of sequence motifs of defined length (Fig. 2b). Further quality checks can be done to assess the diversity (uniqueness) of full sequences as well as to assess if the random region length reflects ones expectations (Fig. 2c). Library quality checks are typically performed on the basis of millions of sequences and enable scientists to assess if a respective library is suitable to be used as a high diverse source for potential aptamer ligands in subsequent laborious SELEX experiments. Moreover, iterative rounds of quality control of NGS data and corresponding adjustments of different library synthesis parameters enable the establishment of protocols for reliable high quality starting library synthesis (Fig. 2a–c).

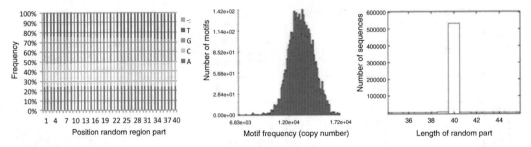

Fig. 2 Quality control result types for an equally random distributed starting library with a 40 nt long random region. (**a**) Distribution of nucleotide building blocks. An equally random distributed starting library contains about 25 % of A, C, G, and T at each position. (**b**) Distribution of sequence motifs of defined length. The Gaussion distribution visualizes an equal random distribution also on the level of motifs (here of length six). (**c**) Check for random region length distribution enables to assess if the length of the random part is consistent with the expectations. Almost all sequences have the expected length of 40 nt in this example

2.3 Improved Aptamer Identification

NGS data derived from enriched SELEX libraries enable an insight into the black box of in vitro selection experiments. Values on various levels can be created, including (1) the assessment of the efficiency of the applied SELEX protocol, (2) identification of aptamers in very early selection rounds, (3) identification of rare aptamers, (4) identification of aptamers addressing defined target epitopes, and (5) identification of aptamers in the context of sequence families.

2.3.1 Assessing the Efficiency of In Vitro Selection

The comparison of sequence diversity of successive selection rounds allows to evaluate the efficiency of a SELEX experiment: The decrease of unique sequences in the course of successive SELEX cycles can be used as an indicator for sequence enrichment (Fig. 3a). COMPAS also provides results that enable the assessment of sequence enrichment at high resolution (Fig. 3b): The number as well as the amount of individual clones which are enriched in the course of an experiment is revealed at a high resolution and allows to asses the influence and efficiency of applied selection conditions, e.g., the stringencies in separation steps for the removal of unbound/weak binders by addition of competitors, more washing, and variation of target-pool incubation times to select for small on rates or high off rates. The results enable the scientist to learn how SELEX really works and consequently to establish more efficient SELEX protocols.

2.3.2 Identification of Aptamers at High Resolution

NGS and subsequent data analysis of enriched libraries enables the identification of hundreds to thousands of potential aptamer candidates. The most frequent sequences are not automatically those with the expected binding properties. For this reason, a strategy which allows the efficient deconvolution of potential aptamer ligands gets mandatory. It enables the scientist to decide which

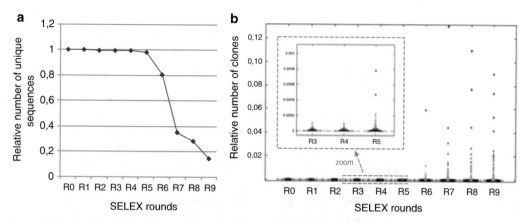

Fig. 3 Result types enabling to assess the efficiency of SELEX experiments. (**a**) Tracking the percentage of unique sequences in the course of SELEX experiments. All sequences are unique in the starting library (R0). The number of unique sequences continuously decreases as a function of enrichment of individual potential aptamers during successive SELEX rounds one to nine (R1–R9). Clear enrichment takes place from round five (R5) to round six (R6) first time where the number of unique sequences decreases from 98 % to 80 %. (**b**) Enrichment of sequences in the course of SELEX rounds R1–R9 is visualized on a monoclonal level. The diagram depicts the number of individual sequences of each round thereby enabling the scientist to analyze sequence enrichment as well as sequence diversity in each round. Zooming enables to asses the development of potential aptamers at very high resolution

representative sequences are synthesized and subsequently tested in binding or functional assays. The COMPAS software applies a deconvolution procedure within two steps:

Deconvolution step "one." In this step millions of aptamer sequences of an enriched SELEX library are clustered by the criteria of similarity into sequence families. Dependent on the amount of enrichment and applied similarity parameters hundred to several hundreds of sequence families are usually identified. Clustering can be performed by applying different methodologies: the most simplest is the counting and ranking of the most frequent identical full length sequences followed by fuzzy searching for other sequences that differ by a defined number of nucleotides building blocks. However, in very early selection rounds where enrichment is not clear and not yet reflected by the presence of multiple copies of full sequences such a methodology does not comprehensively and reliably work [7]. A second method termed "co-occurrence method" was exclusively developed for aptamer identification in very early rounds: first, shorter sequence parts (motifs or k-mers, typically of length 4–8) are identified and ranked. Subsequently COMPAS searches for sequences where two of the identified frequent motifs co-occur. Within several iterative rounds of proto-pattern-formation finally sequence families are clustered.

Sequences belong to a particular family when they contain two common subsequences (a.k.a. motifs) and their remaining segments demonstrate marked similarity. A third method clusters sequence families by the criteria of Shannon's information entropy. The membership of sequences to families is successively tested while monitoring their influence on the family's relative information entropy—a family is defined via the first minimum. Also a combination of these methods is possible to finally yield sequence families. Dependent on amount of enrichment and the applied sequencing depth each family may be build up by tens (early rounds) to several hundreds (later rounds) of closely related sequences. Independent from the applied clustering algorithm and independent from the size of each family, one family is represented by one aptamer in the first instance. This aptamer representing a family can be synthesized and tested in binding- or functional assays. COMPAS lists sequence families in the order of frequency. For each family a consensus sequence including its absolute and relative number is given. Sequences of family members including absolute and relative numbers are also listed (Fig. 4a). Each sequence family is finally visualized in a diagram showing the distribution of nucleotide bases over the random region positions (Fig. 4b). With respect to potential aptamer optimization the computed diagrams enable scientists to get first ideas concerning constant and variable sequence positions.

Deconvolution step "two"—clustering of aptamer families into family clans. As a second measure for systematic reduction of number of sequences to be synthesized and tested, an additional deconvolution step can be applied. Within this second step each consensus sequence representing an aptamer family is analyzed by applying alternative similarity criteria. Such criteria can for example be the occurrence of the same but differently arranged sequence parts or the occurrence of the same potential loop motifs. COMPAS sorts the consensus sequences that have been generated in deconvolution step "one" on the basis of these similarity criteria and computes a heatmap that visualizes clusters of aptamer families that share high similarities. In COMPAS we term such clusters a "sequence family clan." Once an active family clan has been identified (e.g., containing aptamers addressing the functional site of a protein), the aptamer families and family members building up this clan can be examined and tested more closely.

An alternative variant for deconvolution step "two" is the tracing of the frequency of one representative sequence for each aptamer family over other SELEX rounds. This strategy opens up new avenues in the fields of aptamer identification, as it enables to compare SELEX rounds on the level of monoclonal sequences. SELEX experiments can be designed in a way that additional control-SELEX experiments can be performed in parallel to SELEX

a

Name	SEQUENCE	Abs. No.	Rel. No.
F9M1	TAGAGGAGCACCCAACAACCTTAACTGAACTACACTGCTTCTTGAAA	10	2.695e-05
F9M2	TAGAGGAGCACCCAACAACCTTAACTGAATTACACTGCTTCTTGAAA	5	1.348e-05
F9M3	TAGAGGAGCACCCAACAACCATAACTGAATTACACTGCTTCTTGAAA	2	5.390e-06
F9M4	TAGAGGAGCACCCAACAACCTTAACTGAATTACACTGCTTCTTGAAAA	1	2.695e-06
F9M5	TGGAGGAGCACCCAACAACCTTAACTGAACTACACTGCTTCTTGAAA	1	2.695e-06
F9M6	TAGAGGAGCACCCAACAACCTTAACTGAATTACACTGCTTCTAAAAA	1	2.695e-06
F9M7	TAGAGGAGCACCCAACAACCTTAGCTGAATTACACTGCTTCTTGAAA	1	2.695e-06
F9CONS	**TAGAGGAGCACCCAACAACCTTAACTGAACTACACTGCTTCTTGAAA**	**21**	**5.660e-05**

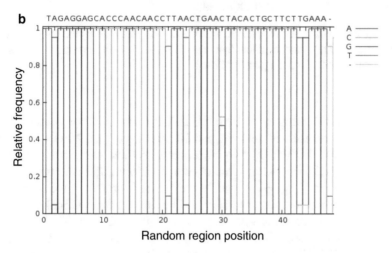

Fig. 4 Example of aptamer sequence family "F9CONS." (**a**) The family is composed out of seven different monoclonal family members (F9M1–F9M7) and comprises 21 sequences. For individual family members as well as for the complete family absolute and relative numbers are listed. The given consensus sequence (F9CONS) represents the respective aptamer family. (**b**) Graphical overview of the random part of family F9CONS: the diagram visualizes which nucleotides are fixed at a defined position and which positions are allowed to vary within the given sequence family

against the target of interest which allows to monitor if (or to which extend) defined monoclonal sequences are present or absent in this control-SELEX experiments. Such control-SELEX experiments can be (1) performed against closely related targets, or (2) in the presence of a specific competitor or (3) against cell lines expressing the target on the surface. Dependent on the type and nature of control-SELEX experiments the scientist can hypothesize if aptamer ligands of interest can be expected to be enriched or decreased in respective control experiments. In doing so, aptamers targeting defined binding sites can already be predicted in silico (Fig. 5). Blank et al. applied this strategy to identify an aptamer addressing the target gp120 at its interaction site to the

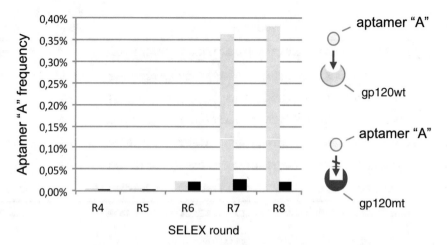

Fig. 5 Tracking the frequency of aptamer "A" over SELEX rounds performed against the target gp120wt (*light grey bars*) and gp120mt (*dark grey bars*), a variant that was mutated at the binding side of interest. Aptamer "A" shows clear and exclusive enrichment against gp120wt

CD4 receptor [2]. Six SELEX rounds were performed against the recombinant protein gp120 wild type (gp120wt). Then SELEX was branched: two additional rounds were performed against gp120wt and in parallel two additional SELEX rounds were performed against a gp120 variant that was mutated at the binding site of interest (gp120mt). The SELEX rounds against gp120wt as well as against gp120mt were digitalized by NGS. In the next step the relative frequency of aptamer sequences identified in gp120wt SELEX were monitored over NGS datasets derived from gp120wt- as well as from gp120mt SELEX. The hypothesis was that the aptamers of interest get exclusively enriched against the wild-type protein. Finally one aptamer could be predicted in silico which showed continuous enrichment in gp120wt SELEX, but was not further enriched in rounds seven and eight that were performed against the mutated protein. Radioactive filter retention assays against gp120wt as well as gp120mt confirmed exclusive binding to the wild-type form.

The two-step deconvolution strategy predominantly serves the purpose to filter the high number of potential aptamer candidates for those with the desired binding properties. Other clustering strategies for deconvolution step "two" are in various stages of development. Amongst them a method that considers aptamer structures and therefore builds up family clans by the criterion of high similarity of common potential sequence loops. However, as deconvolution steps at the end cluster sequences with defined similarities, the clustered data may also be used to learn more about specific sequences in order to optimize them.

2.4 Lead Optimization

To increase the aptamers half-life in biological fluids against degradation by nucleases, scientist iteratively replace all possible natural nucleotide building blocks against unnatural building blocks and test weather the modified aptamers binding properties have been at least maintained. Additionally, for an economic production of aptamers the typically 80–100 nt long macromolecules have to be truncated down to a size of about 40 nt. Post-SELEX optimizations are laborious and any information that enables scientists to assess the role of defined aptamer regions (sequence motifs that mediate target binding; sequence motifs that build the aptamer scaffold) will streamline the optimization process.

The COMPAS software supports this process: As soon as aptamers with the desired binding properties have been identified, the sequence space of its family, family clan, or group of sequences with similar enrichment behavior can be analyzed in respect to the identification of common sequence parts as kind of lowest common denominator which then may indicate motifs or motif combinations which are essential for binding. Other, non-conserved sequence positions can be analyzed in the light of being less critical for binding. Maybe they are part of a stabilizing stem that is accessible for truncation or the respective nucleotide can be tested weather it is possible to replace it against an unnatural building block to stabilize the aptamer against enzymatic degradation.

3 Outlook

In silico analysis of NGS datasets derived from starting as well as from enriched SELEX rounds provide value on several stages (Fig. 1). Additional analysis upgrading and complementing the traditional SELEX process can be performed and results can be generated that finally accelerate aptamer identification and optimization. Moreover, in silico analysis offers dynamic insight into selection processes. It can be expected that NGS and in silico analysis will contribute to a profound understanding of the influence of applied selection pressures. This will enable to further streamline in vitro evolution processes on different platforms. In contrast to genomic applications that rely on the analysis on NGS data and for which a number of well-established software solutions already exist, user-friendly tools in the field of SELEX are on various stages of development [10, 11] but not readily available. Software development has to consider that experts in the field of SELEX are not automatically experienced in the field of computational data analysis. For this reason software tools that are dedicated to really contribute to a better aptamer identification have to meet a variety of requirements: First, a convenient usability of the software tool—for the parsing of NGS data as well as for the straightforward generation of results in all the different application fields (Fig. 1) is mandatory. Second, a graphical user-interface that enables an intuitive handling and analysis of sequence data has to be present for a flexible and wide applicability. Algorithms have to be applied that are applicable for computing millions of sequences in a reasonable time on standard computers. These algorithms should at the same time function without the error-prone query for parameters. Comprehensible graphics and tables have to be generated to enable scientists the easy interpretation of a SELEX experiment. Ideally, the graphs are zoomable and interactive to extract relevant information also at high resolution. Additionally, the possibility to annotate enriched sequences with user defined descriptions by assigning user defined databases helps researches to identify tag, and separation matrix binding sequences or possible contaminations.

COMPAS fulfills these requirements. It provides a method to scientists without noteworthy computer skills to extract relevant information from all stages of the SELEX process: Finally a more economic identification of aptamers with the expected properties gets possible. As in silico analysis also enables to understand the influence of applied selection conditions, every new analyzed experiment successively brings more light into the black box of in vitro selection.

References

1. Kahvejian A, Quackenbush J, Thompson JF (2008) What would you do if you could sequence everything? Nat Biotechnol 26:1125–1133

2. Blind M, Blank M (2015) Aptamer selection technology and recent advances. Mol Ther Nucleic Acids 4:e223

3. Ditzler MA, Lange MJ, Bose D, Bottoms CA, Virkler KF, Sawyer AW et al (2013) High-throughput sequence analysis reveals structural diversity and improved potency among RNA inhibitors of HIV reverse transcriptase. Nucleic Acids Res 41:1873–1884

4. Cho M, Soh HT (2010) Quantitative selection of DNA aptamers through microfluidic selection and high-throughput sequencing. Proc Natl Acad Sci U S A 107:15373–15378

5. Kupakuwana GV, Crill JE, McPike MP, Borer PN (2011) Acyclic identification of aptamers for human alpha-thrombin using over-represented libraries and deep sequencing. PLoS One 6, e19395

6. Schütze T, Wilhelm B, Greiner N, Braun H, Peter F, Mörl M et al (2011) Probing the SELEX process with next-generation sequencing. PLoS One 6:e29604

7. Hoon S, Zhou B, Janda K, Brenner S, Scolnick J (2011) Aptamer selection by high-throughput sequencing and informatic analysis. Biotechniques 51:413–416

8. Thiel KW, Hernandez LI, Dassie JP, Thiel WH, Liu X, Stockdale KR et al (2012) Delivery of chemo-sensitizing siRNAs to HER2+-breast cancer cells using RNA aptamers. Nucleic Acids Res 40:6319–6337

9. Berezhnoy A, Stewart CA, Mcnamara JO II, Thiel W, Giangrande P, Trinchieri G et al (2009) Isolation and optimization of murine IL-10 receptor blocking oligonucleotide aptamers using high-throughput sequencing. Mol Ther 20:1242–1250

10. Hoinka J, Berezhnoy A, Dao P, Sauna ZE, Gilboa E, Przytycka TM (2015) Large scale analysis of mutational landscape in HT-SELEX improves aptamer discovery. Nucleic Acids Res. doi:10.1093/nar/gkv308

11. Alam KK, Chang JL, Burke DH (2015) FASTaptamer: a bioinformatic toolkit for high-throughput sequence analysis of combinatorial selections. Mol Ther Nucleic Acids 4:e230

12. Levay A, Brennemann R, Hoinka J, Sant D, Cardone M, Trinchieri G et al (2015) Nucleic Acids Res. doi: 10.1093/nar/gkv534. Epub 2015 May 24

13. Dias-Neto E, Nunes DN, Giordano RJ, Sun J, Botz GH, Yang K et al (2009) Next-generation

phage display: integrating and comparing available molecular tools to enable cost-effective high-throughput analysis. PLoS One 4:e8338

14. Ravn U, Gueneau F, Baerlocher L, Osteras M, Desmurs M, Malinge P et al (2010) By-passing in vitro screening—next generation sequencing technologies applied to antibody display and in silico candidate selection. Nucleic Acids Res 38:e193

15. 't Hoen PA, Jirka SM, Ten Broeke BR, Schultes EA, Aguilera B, Pang KH et al (2012) Phage display screening without repetitious selection rounds. Anal Biochem 421:622–631

16. Mannocci L, Zhang Y, Scheuermann J, Leimbacher M, De Bellis G, Rizzi E et al (2008) High-throughput sequencing allows the identification of binding molecules isolated from DNA-encoded chemical libraries. Proc Natl Acad Sci U S A 105:17670–17675

Part II

Characterization

Chapter 8

Aptamer Binding Studies Using MicroScale Thermophoresis

Dennis Breitsprecher, Nina Schlinck, David Witte, Stefan Duhr, Philipp Baaske, and Thomas Schubert

Abstract

The characterization and development of highly specific aptamers requires the analysis of the interaction strength between aptamer and target. MicroScale Thermophoresis (MST) is a rapid and precise method to quantify biomolecular interactions in solution at microliter scale. The basis of this technology is a physical effect referred to as thermophoresis, which describes the directed movement of molecules through temperature gradients. The thermophoretic properties of a molecule depend on its size, charge, and hydration shell. Since at least one of these parameters is altered upon binding of a ligand, this method can be used to analyze virtually any biomolecular interaction in any buffer or complex bioliquid. This section provides a detailed protocol describing how MST is used to obtain quantitative binding parameters for aptamer–target interactions. The two DNA-aptamers HD1 and HD22, which are targeted against human thrombin, are used as model systems to demonstrate a rapid and straightforward screening approach to determine optimal buffer conditions.

Key words MicroScale, Thermophoresis, Binding assay, Dissociation constant, Aptamer–target interactions, Interaction affinity

1 Introduction

The multifunctional serine protease thrombin plays a key role in coagulation, inflammation, and cellular proliferation events [1]. The enzyme represents an important therapeutic target due to its functions in the conversion of fibrinogen to fibrin and concomitant platelet aggregation. In the past, several DNA and RNA aptamers were selected against thrombin with the aim to inhibit enzyme activity. The 15 bp Thrombin Binding Aptamer TBA (also named HD1 or G15D) was the first to be described blocking the procoagulant functions of thrombin [2]. This DNA-aptamer forms a G-quadruplex structure and binds to the exosite I of thrombin, which represents the binding site for fibrinogen and other substrates [3]. A second quadruplex forming DNA-aptamer named 60–18

Günter Mayer (ed.), *Nucleic Acid Aptamers: Selection, Characterization, and Application*, Methods in Molecular Biology, vol. 1380, DOI 10.1007/978-1-4939-3197-2_8, © Springer Science+Business Media New York 2016

(29) or HD22 was shown to bind to the second anion binding site, the exosite II. Glycosaminoglycans like heparin and other ligands interact with thrombin via this site [4]. In further studies, additional RNA and DNA aptamers to thrombin were selected and characterized [5, 6].

Essential for the characterisation and development of potential therapeutic aptamer candidates is the analysis of basic binding parameters. MicroScale Thermophoresis (MST) represents a powerful technology, which is well suited to quantify affinities of aptamer–target interactions.

The basic principal behind MicroScale Thermophoresis is a physical effect referred to as thermophoresis [7–10]: This phenomenon describes the directed movement of molecules through temperature gradients depending on their size, charge, and hydration shell. Upon binding of a ligand to a molecule, at least one of these parameters is changed, resulting in an altered thermophoretic movement of the unbound and bound states [11].

The MicroScale Thermophoresis technology monitors the directed movement of molecules through μm-sized temperature gradients by detecting the fluorescence of the labeled molecule. The schematic setup of a Monolith™ instrument is shown in Fig. 1a.

MST Experiments are carried out in aqueous buffer in capillaries with a volume of 4–6 μl. An infrared laser establishes a microscopic temperature difference ΔT, which spans 2–6 °C, depending on the instrument settings, and is restricted to a volume with a diameter of ~50 μm. This results in thermophoretic depletion or accumulation of molecules in the region of elevated temperature, which is quantified by the Soret coefficient

$$S_T : c_{hot} / c_{cold} = exp(-S_T \Delta T).$$

Each capillary is successively scanned and the movement of fluorescent molecules through the temperature gradient is monitored by fluorescence detection through an objective.

A typical MST experiment is shown in Fig. 1b. Initially, fluorescence in the sample is detected in the absence of a temperature gradient to ensure homogeneity of the sample. After 5 s, the IR-laser is activated leading to the establishment of the temperature gradient. This causes an initial steep drop of the fluorescence signal—the so-called Temperature- or T-Jump—which reflects the temperature dependence of the fluorophore quantum yield. After the T-Jump, a slower thermophoresis-driven depletion of fluorophores occurs. Once the IR-laser is deactivated, a reverse T-Jump and subsequent backdiffusion of fluorescent molecules can be observed. Please note that both the T-Jump as well as the thermophoresis signal can display an increase or a decrease in fluorescence intensity. In case of thermophoresis, movement of molecules towards regions with lower temperature is referred to as positive thermophoresis, while

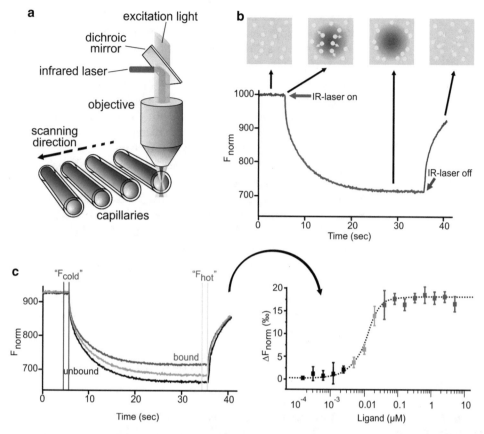

Fig. 1 MicroScale Thermophoresis binding assays: (**a**) Schematic representation of the technical setup of a Monolith™ instrument used to monitor MST. (**b**) Plot of fluorescence intensity changes over time throughout a typical MST experiment. The schematic representations depict movement of fluorescent molecules through the temperature gradient. (**c**) Results of a MST-based binding experiment. The bound and unbound states of the molecule display different thermophoretic depletion profiles. The difference in thermophoresis can be plotted as a change in normalized fluorescence versus ligand concentration to obtain a binding curve. The *dashed line* corresponds to the best fit calculated based on the law of mass action

movement towards regions with higher temperature is referred to as negative thermophoresis.

Since thermophoresis is highly sensitive towards changes in molecular properties, a binding event induces changes in thermophoresis, which can be used to determine equilibrium constants, such as the dissociation constant K_d. For this, a serial dilution of the ligand is prepared, mixed with a constant concentration of labeled target molecule, loaded into capillaries and analyzed in the instrument by subsequent scanning of each capillary. The changes in thermophoresis are then plotted and used to derive the binding constant. The results of a typical binding experiment are illustrated in Fig. 1c.

Here we provide a detailed protocol for MST-based analysis of aptamer binding: Generally, for binding experiments, the Cy5-labeled aptamer (Here: HD1 or HD22) is kept at a constant concentration in each capillary, and the concentration of the unlabelled molecule (here: thrombin) is stepwise reduced by a 16-fold 1:1 serial dilution. Changes in aptamer thermophoresis upon thrombin binding are measured and plotted as a function of the thrombin concentration, and the dissociation constant is calculated. We moreover show how the affinity of the HD22 aptamer–thrombin interaction changes depending on buffer conditions across different pH values and salt concentrations. Furthermore, the binding of HD1 and HD22 to thrombin is analyzed under close physiological conditions in serum. The results indicate that the binding between HD22 and thrombin is highly dependent on the salt conditions, whereas changes in the pH show only minor effects on the binding properties of the aptamer. Interestingly, MST measurements in serum demonstrate a considerable influence on the binding affinity of HD22, whereas the binding parameters of HD1 are only changed to a minor extent, highlighting the importance of aptamer binding studies under close-to-native conditions.

2   Materials

2.1   Buffers and Reaction Partners

1. Aptamer binding buffer: 50 mM Tris–HCl, pH 7.6, 100 mM NaCl, 1 mM $MgCl_2$, 0.05 % Tween 20.

    In order to analyze buffer dependencies of the binding parameters, the composition of the binding buffer was varied. The effect of different NaCl concentrations on the thrombin–HD22 interaction was determined by stepwise increasing the NaCl concentrations from 0 mM to 1000 mM. In addition, the dependency of aptamer binding properties on the pH was analyzed in a range from pH 6.8–9.2. In order to analyze the binding properties of HD1 and HD22 in serum, fetal calf serum was used as buffer.

2. Fluorescently labeled HD1 and HD22 aptamers (HD1: 5′-Cy5-GGTTGGTGTGGTTGG-3′, HD22: 5′-Cy5-AGTC CGTGGTAGGGCAGGTTGGGGTGACT-3′).   The oligonucleotides were dissolved in water (100 µM final concentration) and stored in light protected vials at −20 °C.

3. Human thrombin alpha (Haematologic Technologies Inc., Essex Junction, Vermont) as a 220 µM stock solution and stored at −20 °C.

2.2   Microscale Thermophoresis Equipment

1. Microscale Thermophoresis instrument Monolith NT.115[Pico] (NanoTemper Technologies, Munich, Germany).

2. Monolith NT™ capillaries (NanoTemper Technologies, Standard treated, Hydrophobic and Hydrophilic).

3   Methods

3.1   Preparing the Aptamer Working Stock

To determine dissociation constants from a serial dilution, the concentration of the fluorescently labeled molecule should be close to or below the expected K_d. For optimal results, the concentration of the fluorescently labeled molecule and the LED should be adjusted in such a way that the detected fluorescence intensity lies between 2500 and 18,000 fluorescence counts on the Monolith NT 115[Pico]. Low excitation intensities (low LED powers) are preferable to minimize photobleaching of the sample. For more info please (*see* **Note 1**). Typically, an aptamer–thrombin binding reaction contains a final concentration of 1 nM fluorescently labeled aptamer with varying thrombin concentrations.

1. Dissolve oligonucleotides according to the manufacturer's instructions and determine the exact nucleic acid concentration using a UV/VIS Spectrophotometer.

2. Prepare the aptamer working stock by diluting the oligonucleotide stock to 500 nM with the respective buffer. Incubate the mixture for 2 min at 90 °C, then let the samples slowly cool down until they reach room temperature. The sample can now be stored at 4 °C for 2 weeks. Dilute the working stock further down to 2 nM in 180 μl reaction buffer or FCS.

3.2   Preparation of the Thrombin Titration Series

A MST titration series contains up to 16 capillaries, which are measured in a single thermophoresis run. Dilutions of the unlabelled thrombin should start at a concentration, which is about 40-fold higher than the expected K_d.

1. Pre-dilute the 220 μM thrombin stock with the respective buffer or with FCS to 2000 nM for the HD22 experiments in different buffers and to 10,000 nM for the serum measurements. Prepare a 16-step 1:1 (v/v) serial dilution of thrombin in the respective binding buffer, so that each dilution step reduces the protein concentration by 50 %. For this, 16 small micro reaction tubes should be prepared: Label the reaction tubes from 1 through 16. Fill 20 μl of a 2000 nM thrombin (for experiments with HD22 in different buffers) or 10,000 nM thrombin (for experiments with HD1) in the respective binding buffer or FCS in tube 1.

2. Add 10 μl of aptamer binding buffer into the micro reaction tubes 2–16. Transfer 10 μl of tube 1 to tube 2 and mix very well by pipetting up and down several times, transfer 10 μl to the next tube and repeat this dilution for the remaining tubes.

It is important to avoid any buffer dilution effects. The buffer in tube 1 and the buffer in the tubes 2–16 must be identical. Protein low binding tubes are recommended since thrombin tends to adsorb to surfaces. Please note, that the NanoTemper analysis software contains a function (concentration finder) that can be used to determine the optimal concentration range of the titration partner (*see* **Note 2**).

3.3 Preparation of the Final Reaction Mix

The individual binding reactions should be prepared with an optimal volume of 20 μl (10 μl aptamer working solution + 10 μl of the respective thrombin dilution), for the ease of pipetting and the minimization of experimental errors. However, a volume of only 4 μl is sufficient to fill the capillary.

1. Add 10 μl of the 2 nM aptamer working solution to 10 μl of each thrombin dilution (*see* Subheading 3.2). Mix the sample by pipetting and briefly centrifuge the samples. Consider this initial dilution step when calculating the final concentrations of thrombin and aptamer.

2. Incubate the samples for 5 min and fill the samples into hydrophilic capillaries (*see* **Note 3**). For thrombin, hydrophilic capillaries are recommended to avoid adsorption to glass capillaries.

The capillaries are placed onto the capillary holder tray, which is then inserted into the Monolith NT.115^{Pico} instrument. Use the NT Control software to set up experiment parameters and start the MST experiment (*see* below).

3.4 Capillary Scan

Prior to the MST measurement, the position of each capillary is automatically determined in an initial capillary scan. The capillary scan also provides important information about sample quality: Adsorption of fluorescent molecules to the capillary walls can easily be detected (*see* **Note 3**), and pipetting errors or fluorescence quenching effects can be identified prior to the experiment. Hence, the capillary scan also serves as an important quality control to quickly identify irregularities and to accordingly optimize the glass capillary type, buffer condition or sample quality.

1. After starting the NT Control software, make sure that the LED channel is set to "red" for Cy5-dyes (Note: while regular NT.115 instruments have two LED channels of different wavelengths—a combination of blue, green or red—the NT.115^{Pico} has only one red LED channel). Start the capillary scan by clicking on the respective button "start capillary scan" (initial settings: LED Power: 20 %).

The fluorescence signal during the capillary scan should be between 2500 and 18,000 fluorescence units (Monolith NT.115^{Pico}). If the value is below 2500 fluorescence units, please refer to (*see* **Note 4**). Since all samples should contain the same concentration of fluorescently labeled DNA, individual differences in intensity between capillaries should be below 10 % (*see* **Note 5** for trouble shooting if the variance in overall fluorescence is larger).

3.5 MST Measurement

After completing the capillary scan, the MST-measurement can be performed.

1. First, assign the thrombin concentrations from the dilution series to the respective capillary position. For this, enter the highest concentration of thrombin (final 1000 nM or 500 nM, respectively) for capillary 1, select the correct dilution type (in this case 1:1), click on the maximum concentration and use the implemented drag-function to automatically add the remaining concentrations in capillaries #2–16.

2. Enter the concentration of the fluorescent aptamer (final 1 nM).

3. NanoTemper Monolith™ devices are temperature controlled in a range from 22 to 45 °C. To select the desired reaction temperature, enable and activate the temperature control. For typical experiments the temperature is kept constant at 25 °C.

4. The MST power is adjusted to 80 %, as previous experiments showed an optimal MST signal under these conditions (*see* **Note 6**). Keep in mind that the strength of the temperature gradient induced by the IR-laser correlates with the MST power. For more information on the MST power *see* **Note 7**.

5. When using default settings, the fluorescence is detected for 5 s, and then MST is recorded for 30 s. After inactivation of MST, fluorescence is recorded for another 5 s, monitoring backdiffusion of molecules. For high MST powers ≥ 80 %, we recommend reducing the "MST ON" time to <15 s.

6. After selecting the destination folder to save the experiment, start the MST measurement by pressing the start button. Using the above-mentioned settings, one measurement will be completed in about 10–15 min.

7. Repeat the measurements two more times for a more accurate determination of the K_d value.

3.6 MST Data Analysis

Data can be analyzed already during data acquisition (on the fly data analysis) with the provided NT-Analysis software.

A plot of typical MST traces and a plot of the changes in the normalized fluorescence (ΔF_{norm}) versus the ligand concentration are shown in Fig. 1b, c. Both plots are important for data analysis. Inspection of the normalized MST traces offers additional information on aggregation and precipitation effects, thus representing another important quality control feature of the MST technology (*see* **Note 8**).

In order to determine the binding constants from the ligand-induced changes in thermophoresis, the ligand-dependent changes in normalized fluorescence F_{norm} are calculated with

$$F_{\text{norm}} = F_{(\text{hot})} / F_{(cold)}$$

where $F_{(hot)}$ and $F_{(cold)}$ represent averaged fluorescence intensities at defined time points of the MST traces. By default, three different cursor-settings can be chosen to analyze the data: Thermophoresis, Thermophoresis + T-Jump, and T-Jump only. For information on the different settings (*see* **Note 9**). Once F_{norm} for the chosen cursor settings is plotted, the data can be fitted to obtain either the dissociation constant (K_d) (Fig. 1c; *right*) or the EC_{50} value (*see* **Note 9**).

Step-by-step data analysis:

1. Import the acquired data using the "load project" button and then select the data set. Select the MST runs you want to analyze.

2. Choose either Thermophoresis, Thermophoresis and T-jump, or T-jump cursor settings. The respective F_{norm} values will be plotted. For the analysis of aptamer–thrombin data, select the T-Jump cursor settings.

3. The dissociation constant K_d of the interaction is determined by fitting the data using MST-standard fit algorithms (law of mass action, *see* **Note 9**).

   F_{norm} values and the corresponding fit can be exported as a text or excel file, and results can also be summarized as a report in pdf format. In the presented set of experiments, the influence of the salt conditions and of the pH value on the aptamer–thrombin affinity can be quantified by comparing the binding isotherms and K_d values of the different experiments (Fig. 2a, d).

4. For a better side-by-side comparison, the data can be normalized (Fig. 2b, e) to the fraction of bound molecules (FB) by the following equation:

$$FB = (value(c) - free) / (complexed - free),$$

In which *value(c)* is the MST-value measured for the concentration *c*, *free* is the MST-value for the unbound state (lowest concentration) and *complexed* is the MST-value for the fully bound state.

In this study, the influence of salt and pH conditions on the binding parameters of HD22 to thrombin was determined by MicroScale Thermophoresis. The affinity was found to strongly depend on salt concentrations (Fig. 2a–c) whereas the pH had only minor effects on affinity (Fig. 2d–f). In a further experiment HD22 and HD1 aptamers towards thrombin were tested directly in serum. Importantly, while the affinity of HD22 was decreased by a factor of ~100 in serum when compared to buffer (Fig. 2g, i), the affinity of HD1 was nearly unchanged (Fig. 2h, i), suggesting that HD22 also interacts with other components in serum. These results demonstrate that MicroScale Thermophoresis is an

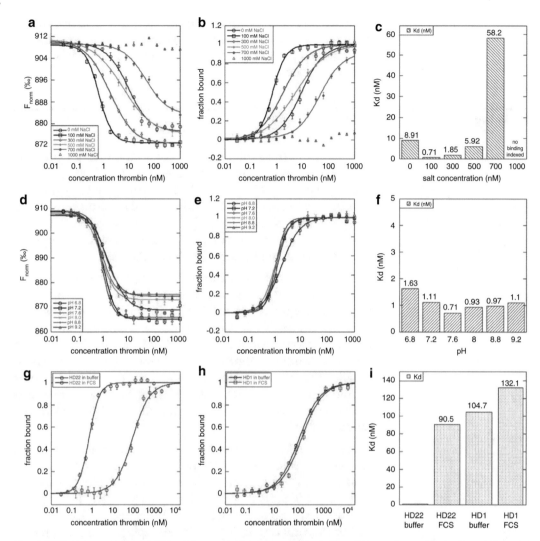

Fig. 2 MST data analysis: (**a**) Plot of the normalized fluorescence F_{norm} (‰) vs. the concentration of thrombin from MST experiments. Data were derived from the T-Jump signal. NaCl concentration was varied between 0 and 1000 mM. *Lines* represent fits of the data points using the law of mass action. (**b**) Fraction bound plot of the data shown in (**a**). (**c**) Quantitative comparison of the NaCl-dependence of the K_d of the thrombin–aptamer interaction. The interaction shows a strong salt-dependence, with optimal binding at 100 mM NaCl. (**d**) Plot of the normalized fluorescence F_{norm} (‰) vs. the concentration of thrombin from MST experiments. Data derived from the T-Jump signal. The pH of the buffer was varied between 6.8 and 9.2. *Lines* represent fits of the data points using the law of mass action. (**e**) Fraction bound plot of the data shown in (**d**). (**f**) Quantitative comparison of the pH-dependence of the K_d of the thrombin–aptamer interaction. pH dependence is generally low, with the strongest interaction occurring at a pH value of 7.6. (**g**) Fraction bound plot of the MST experiments HD22 aptamer binding to thrombin in buffer or serum. (**h**) Fraction bound plot of the MST experiments HD1 aptamer binding to thrombin in buffer or serum. (**i**) Quantitative comparison of the buffer-dependence of the K_d of the thrombin–aptamer interaction. Serum has a considerable influence on the binding of HD22 to thrombin. The affinity decreases from 0.71 nM to 90.5 nM

easy-to-use, fast and precise method to study aptamer–target interactions. MST also allows a free choice of buffer conditions, even in close-to physiological environment such as serum, whole blood or cell lysate.

4 Notes

1. The detection of low fluorophore concentrations $\ll 1$ nM may require high excitation light intensities (LED power > 75 % using the NT.115^{Pico}), which can cause significant photobleaching during the experiment and introduce additional noise to the binding signal. Use the NanoTemper Anti-Photobleaching kit to eliminate photobleaching throughout the experiment and to optimize the binding signal.

2. The concentration finder tool implemented in the NT.Control and NT.Analysis software simulates binding data and helps finding the right concentration range for the dilution series. It is either possible to simulate how the binding curve will look like given a defined K_d, or within a "K_d Interval", in case the binding affinity can only be roughly estimated. Ideally, the concentration of ligand in the dilution series should be chosen in such a way that at least three data points are present in both the bound and unbound plateau of the binding curve.

3. For successful MST experiments it is imperative that all molecules are free in solution. Some biomolecules however tend to stick to glass surfaces. Adsorption of fluorescent molecules to capillary walls can be identified by irregular capillary profiles in the initial capillary scan. Bumpy, flattened or U-shaped capillary profiles indicate adsorption. *See* Fig. 3a for example profiles. To prevent these "sticking effects" several capillary types—each coated with different passivizing chemicals—are available. These capillaries should be tested for their suitability prior to binding experiments (*compare* Fig. 3a). *Please note*, that the capillaries should not be touched in the center to prevent impurities and adverse effects on the glass surface. Simply dipping in the samples will fill capillaries. Since adhering molecules may falsify the measurement, care should be taken that the capillary does not touch the surface of the reaction tube.

4. If the fluorescence intensity in the capillaries is below 2500 counts, increase the concentration of labeled molecule or increase the LED power. If an increased laser power leads to significant photobleaching, please refer to **Note 1**. Fluorophore concentration of ~10 pM Cy5 are still readily detectable in the NT.115^{Pico} instrument.

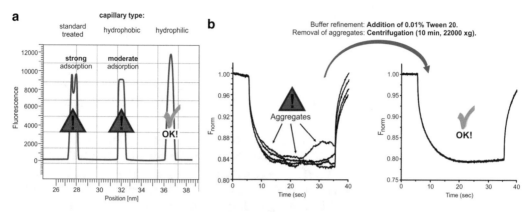

Fig. 3 Assay optimization for MST experiments. (**a**) The capillary scan reveals protein adsorption. Prior to MST experiments, the fluorescent molecule is filled into the different capillary types at the intended final concentration. Molecule-adsorption to capillary walls can be detected by irregular capillary shapes in the capillary scan. Upon adsorption, the fluorescence profile of the capillaries can be U-shaped or flattened. In the example shown here, hydrophilic capillaries prevent molecule adsorption, resulting in a regular capillary shape. (**b**) Detection and removal of aggregates. Aggregated molecules can be identified based on irregular "bumpy" MST traces. Optimization of reaction buffers (e.g., including detergents such as Tween 20 or varying pH values or salt concentration) can improve the solubility of the molecules. Lager aggregates can be efficiently removed by centrifugation. Once aggregation is eliminated, the MST traces of identical samples should be indistinguishable as in the shown example

5. Variations in fluorescence intensities in the capillary scan larger than 10 % can be caused by (a) pipetting errors, (b) aggregation of labeled protein, (c) adsorption to capillary walls (d) fluorescence quenching by the ligand.

   (a) To improve pipetting accuracy, make sure to use the exact same buffer of ligand stock and assay buffer. Mix the solution in each step at least 8 times by pipetting up and down, do not introduce air bubbles while mixing, and do not use the "blow-out" function of you pipette.

   (b) Strategies to minimize aggregation are discussed in **Note 7**.

   (c) Adsorption of capillary walls can be identified and minimized as described in **Note 3**.

   (d) Fluorescence quenching by the ligand results in a systematic rather than a random fluorescence change. To test whether fluorescence loss at increasing ligand concentrations is caused by ligand binding or by ligand-induced denaturation/adsorption of the protein, perform the "SD-Test": Prepare two tubes each containing 10 μl of a 2× SD mix (4 % SDS, 40 mM DTT). Carefully remove 10 μl of tubes 1 and 16 and transfer to the tubes containing the SD-mix, mix well, and incubate for 5 min at 95 °C to denature the protein. Fill both samples into two standard capillaries each and measure the fluorescence intensity. In case of ligand-induced quenching, the fluorescence of denatured protein should be identical for both samples. If you observe a difference in fluorescence intensity for tubes 1 and 16, material was lost either by aggregation and subsequent centrifugation or by unspecific adsorption at the tube walls.

6. For an unknown interaction it is recommended to perform the MST measurement at MST powers of 20 % and 40 %. The lowest MST power which yields clear binding signal is used for evaluation. Only in cases where these settings do not result in clear binding signal higher MST power should be used.

7. A total volume of 2 nl is heated by the infrared laser. The range of the temperature gradient depends on the MST power, and spans 2 °C (MST power 20 %) to 6 °C (MST power 80 %) [9].

8. Aggregation of your protein can be prevented by adding detergents to the assay buffer (0.005–0.1 % Tween 20, 0.01–0.1 % Pluronic F127 or similar), by adding >0.5 mg/mL of stabilizing proteins such as BSA, and/or by centrifugation for >10 min at 22,000 × g prior to the experiment. Aggregates can also be identified by "bumpy" MST traces during the experiment (*see* Fig. 3b).

9. The NanoTemper Analysis software offers two curve fit options: The fit function for K_d from the law of mass action and the fit function for EC_{50} from the Hill equation.

   Kd from law of mass action:

   $$A + T \Leftrightarrow AT$$

   $$Fraction bound = \frac{1}{2c_A}\left(c_T + c_A + K_d - \sqrt{(c_T + c_A + K_d)^2 - 4c_T c_A}\right)$$

   K_d: dissociation constant, to be determined, c_{AT}: concentration of formed complex, c_A: constant concentration of molecule A (fluorescent), known, c_T: concentration of titrated molecule T.

   Please note that the fitting model from the law of mass action describes your data correctly when a molecule A interacts with a molecule B using one binding site or using multiple binding sites with the same affinity.

   EC50 from the Hill equation:

   $$A + nT \Leftrightarrow AT_n$$

   $$Fraction bound = \frac{1}{1 + (EC_{50} / c_T)^n}$$

   c_T: provided concentration of titrated molecule T.

The Hill model can be used to determine the EC_{50} value, which is the concentration of titrant where 50 % of the fluorescent molecule is bound. Please keep in mind that the EC_{50} value is not a physical *constant* like the K_d, but an apparent measure of affinity for one particular experiment, which strongly depends on the used concentrations. The Hill fitting routine should only be used for interactions which are known to be of cooperative nature.

References

1. Siller-Matula JM, Schwameis M, Blann A, Mannhalter C, Jilma B (2011) Thrombin as a multi-functional enzyme. Focus on in vitro and in vivo effects. Thromb Haemost 106(6):1020–1033

2. Bock LC, Griffin LC, Latham JA, Vermaas EH, Toole JJ (1992) Selection of single-stranded DNA molecules that bind and inhibit human thrombin. Nature 355(6360):564–566

3. Macaya RF, Schultze P, Smith FW, Roe JA, Feigon J (1993) Thrombin-binding DNA aptamer forms a unimolecular quadruplex structure in solution. Proc Natl Acad Sci U S A 90(8):3745–3749

4. Tasset DM, Kubik MF, Steiner W (1997) Oligonucleotide inhibitors of human thrombin that bind distinct epitopes. J Mol Biol 272(5):688–698

5. Muller J, Wulffen B, Potzsch B, Mayer G (2007) Multidomain targeting generates a high-affinity thrombin-inhibiting bivalent aptamer. Chembiochem 8(18):2223–2226

6. Becker RC, Povsic T, Cohen MG, Rusconi CP, Sullenger B (2010) Nucleic acid aptamers as antithrombotic agents: opportunities in extracellular therapeutics. Thromb Haemost 103(3):586–595

7. Baaske P, Wienken CJ, Reineck P, Duhr S, Braun D (2010) Optical thermophoresis for quantifying the buffer dependence of aptamer binding. Angew Chem Int Ed Engl 49(12):2238–2241

8. Jerabek-Willemsen M, Wienken CJ, Braun D, Baaske P, Duhr S (2011) Molecular interaction studies using microscale thermophoresis. Assay Drug Dev Technol 9(4):342–353

9. Seidel SA, Dijkman PM, Lea WA, van den Bogaart G, Jerabek-Willemsen M, Lazic A, Joseph JS, Srinivasan P, Baaske P, Simeonov A, Katritch I, Melo FA, Ladbury JE, Schreiber G, Watts A, Braun D, Duhr S (2013) Microscale thermophoresis quantifies biomolecular interactions under previously challenging conditions. Methods 59(3):301–315

10. Zillner K, Jerabek-Willemsen M, Duhr S, Braun D, Langst G, Baaske P (2012) Microscale thermophoresis as a sensitive method to quantify protein: nucleic acid interactions in solution. Methods Mol Biol 815:241–252

11. Ludwig C (1856) Diffusion zwischen ungleich erwärmten Orten gleich zusammengesetzter Lösungen. Sitzungber Bayer Akad Wiss Wien Math-Naturwiss Kl 20:539

Chapter 9

Label-Free Determination of the Dissociation Constant of Small Molecule-Aptamer Interaction by Isothermal Titration Calorimetry

Marc Vogel and Beatrix Suess

Abstract

Isothermal titration calorimetry (ITC) is a powerful label-free technique to determine the binding constant as well as thermodynamic parameters of a binding reaction and is therefore well suited for the analysis of small molecule—RNA aptamer interaction. We will introduce you to the method and present a protocol for sample preparation and the calorimetric measurement. A detailed note section will point out useful tips and pitfalls.

Key words RNA, Aptamer, Small molecule, ITC, K_d, In vitro transcription

1 Introduction

In the last couple of years, aptamers have been introduced as powerful tools in Medicine, Molecular Biology and Synthetic Biology (reviewed in Ref. [1]). The high binding affinity and specificity of aptamers is the prerequisite for many applications; therefore the detailed characterization of their binding behavior is an important step after selection.

A plethora of methods exists to determine the dissociation constant (K_d) of aptamers, including: filter binding assays, fluorescence anisotropy, equilibrium dialysis, surface plasmon resonance (SPR), and differential scanning calorimetry [2]. For most of these methods either the ligand or the aptamer has to be labeled, radioactively or by a fluorescent dye, which may influence the interaction between the aptamer and the ligand and is not always easy to perform. A powerful and label-free method to determine the K_d of aptamers is isothermal titration calorimetry (ITC) [3]. The concentration of aptamer which is needed for one reaction is about 10 µM per ITC experiment, an amount easily obtained by in vitro transcription in any wet lab.

Günter Mayer (ed.), *Nucleic Acid Aptamers: Selection, Characterization, and Application*, Methods in Molecular Biology, vol. 1380, DOI 10.1007/978-1-4939-3197-2_9, © Springer Science+Business Media New York 2016

ITC directly measures the change in the enthalpy (ΔH) caused upon ligand binding. The instrument contains two identical cells, a sample cell and a reference cell. The reference cell is filled with water or buffer and the sample cell contains either the ligand or the aptamer. Defined injections into the sample cell are carried out by a syringe containing the second component. Each injection causes heat effects in the sample cell which is due to dilution effects, stirring and heat changes through the interaction of the different molecules. The amount of power which is needed to compensate these heat changes is directly measured and plotted against time (shown in Fig. 1a).

The energy which is needed until the steady state level is reached again is directly proportional to the energy of the reaction. Since ΔH is directly measured and K_a can be calculated by fitting the ITC binding isotherm, ΔG can be determined using the following equation:

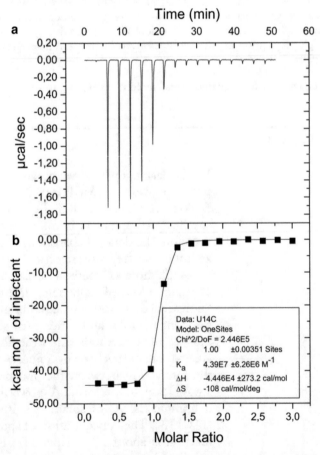

Fig. 1 Example of results obtained by an ITC experiment. The raw data of the experiment are shown in (**a**). Each spike resembles an injection. (**b**) shows the calculated integrals of each spike (*black squares*) and the fit to calculate K_a (*black line*). For fitting the one site binding model was used

$$\Delta G = -RT \ln K_a$$

in which ΔG is the Gibbs free energy, R is the gas constant and T is the temperature in Kelvin. The entropy (ΔS) can also be calculated:

$$\Delta G = \Delta H - T\Delta S$$

Hence, ITC is a powerful technique which determines the K_d as well as other thermodynamic parameters and the stoichiometry of the reaction.

2   Materials

2.1   ITC Instrumentation

A variety of ITC instruments is available on the market. All of them work the same way; they measure the heat changes in the sample cell after the injection of one binding partner. The various instruments are able to measure binding constants (K_a) from 10^3 to 10^9 M^{-1} [2]. A common instrument is the MicroCal iTC200, which is convenient for most applications. In the last couple of years the instruments have become more and more sensitive with the advantage that up to 80 % less sample is needed. The TA Instruments Low Volume Nano ITC, for example, is able to detect as little as 0.05 µJ. Consequently, the sample cell volume can be reduced from 1 ml to 180 µl. Another advantage of low sample instruments is the faster equilibration time, so the injection interval can be shortened. The number of injections also differs between a low volume instrument and a normal volume instrument. With a normal cell instrument about 20–40 injections per experiment are carried out compared to 15 injections with a low volume instrument. As a result, more data points are collected with a normal cell instrument resulting in a better resolution.

All things considered, low volume instruments produce accurate data sets for high enthalpy reactions with binding constants between 10^3 and 10^8 M^{-1}. For low enthalpy reactions instruments with a normal cell volume produce more accurate data.

2.2   Plasmid

1. Recommended plasmid for in vitro transcription: pSP64 Promega. The 3 kbp long plasmid contains a β-lactamase gene as a resistance marker for the selection in E. coli, a multiple cloning site and an origin of replication resulting in multiple copies of plasmid per cell (see Note 1).

2.3   Enzymes and Buffer

1. T7 RNA polymerase for in vitro transcription: The polymerase can be purchased from many suppliers (e.g., New England Biolabs). It can also be purified in the lab following the protocol described by Ref. [4]. In vitro transcription buffer (see Note 2):

250 mM Mg(Ac)$_2$, 1 M Tris–HCl pH 8.0, 1 M DTT, 200 mM spermidine.

2. 10× measurement buffer (*see* **Note 3**): 200 mM Na-cacodylate pH 6.8, 100 mM MgCl$_2$, 1 M NaCl.

---

3 Methods

3.1 General Aspects for In Vitro Transcription

About 5 nmol of RNA is needed for an ITC experiment. The RNA can be chemically or enzymatically synthesized. The enzymatic synthesis by in vitro transcription is a well-established method and can be performed in any lab. For in vitro transcription the sequence of choice needs to be placed on a plasmid downstream of a T7 promoter. At the end of the sequence a restriction site is needed so the plasmid can be linearized by the respective restriction enzyme. During transcription, the T7 RNA polymerase binds to its promoter, transcribes the DNA template into RNA and falls off at the end of the template at the restriction site. This type of transcription is called run-off transcription. The T7 RNA polymerase binds to the template multiple times, this way it produces a high amount of RNA. Thereby, up to several milligrams of RNA can be produced.

3.2 Template Construction for In Vitro Transcription

1. We recommend using the plasmid pSP64 from Promega, a small high copy plasmid. The sequence of choice can easily be inserted into the vector using the restriction sites *Hin*dIII and *Eco* RI of the plasmids multiple cloning site (*see* **Note 4**).

2. The insert should contain the following features: a T7 promoter, the sequence which should be transcribed and a restriction site for run-off transcription (Fig. 2).

3. The T7 promoter sequence is shown in Fig. 2c. The first one of the two Gs at the end will be the first nucleotide of the transcript (*see* **Note 5**).

3.3 3′-End Heterogeneity

1. The T7 RNA polymerase (as well as other polymerases) is known to add non-coded nucleotides to the 3′-end. To avoid this heterogeneity we recommend using a ribozyme for defined 3′ ends. This ribozyme is placed at the 3′ end of the template sequence (Fig. 2a). After the RNA is transcribed the ribozyme cuts the RNA leaving a defined 3′ end.

2. Two ribozymes can be used, the hepatitis delta virus (HDV) ribozyme and the hammerhead ribozyme [5, 6]. For 3′-end heterogeneity the HDV ribozyme is the ribozyme of choice as it has the advantage that it is sequence independently (*see* **Note 6**). The hammerhead ribozyme can also be used, but the sequence needs to be modified depending on the template sequence since the template is part of the hammerhead sequence.

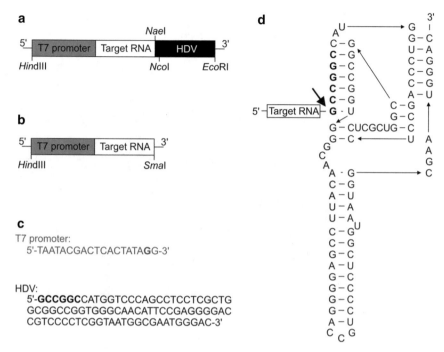

Fig. 2 The constructs for run-off transcription are shown. The T7 promotor is shown in *grey*, the target sequence in white and the HDV ribozyme in black. In (**a**) the construct with HDV ribozyme is shown. The sequence can be cloned into the pSP64 vector by using the restriction sites *Hind*III and *Eco*RI. For run-off transcription the *Eco*RI restriction site can be used. (**b**) shows the construct without the HDV ribozyme. In this case the restriction sites *Hind*III and *Sma*I are used. For run-off transcription the restriction site *Sma*I is used. (**c**) shows the sequence of the T7 promotor with the G in *bold* being the first nucleotide of the transcript. The sequence of the HDV is shown in *black* with the *Nae*I cleavage site indicated in *bold letters*. The secondary structure of the HDV ribozyme is shown in (**d**). The *bold letters* indicate the *Nae*I cleavage site and the *arrow* between the target sequence and the HDV indicates the cleavage site of the ribozyme

3.4 Plasmid Preparation

1. For a large scale in vitro transcription 1–2 mg plasmid is needed (*see* **Notes 7** and **8**).

2. The plasmid pSP64 has to be digested either with the restriction enzyme *Eco*RI, which cleaves the plasmid straight after the HDV ribozyme, or an enzyme which cuts straight after the sequence (*see* **Note 9**).

3. After digestion the plasmid DNA is purified by phenol–chloroform extraction (*see* **Note 10**).

3.5 In Vitro Transcription

1. For the in vitro transcription all components are mixed together as listed in Table 1. It is very important to follow the order given in Table 1; otherwise magnesium phosphate might precipitate.

2. The reaction is mixed by carefully pipetting up and down and is afterwards incubated at 37 °C over night (*see* **Note 11**).

Table 1
In vitro transcription for RNA preparation

Add H₂0 10 ml	RNase-free water
800 μl	Mg(Ac)₂ 250 mM (*see* **Note 12**)
2000 μl	Tris–HCl 1 M pH 8.0
200 μl	Dithiothreitol (DTT) 1 M
100 μl	Spermidine 200 mM
2 mg	Digested plasmid
400 μl	ATP 100 mM
400 μl	UTP 100 mM
400 μl	GTP 100 mM
400 μl	CTP 100 mM
15 μg/ml	T7 RNA polymerase

3.6 HDV Cleavage and RNA Purification

1. The RNA cleaves itself straight after transcription if a ribozyme such as the HDV ribozyme is used. The ribozyme folds into its catalytic active conformation during the incubation at 37 °C. It results in two RNA species, your RNA of interest and the free HDV ribozyme.

2. To eliminate the HDV RNA as well as the plasmid DNA, the in vitro transcription is separated by polyacrylamide gel electrophoresis (*see* **Note 13**). Therefore, the RNA dissolved in deionized formamide containing 25 mM EDTA before it is loaded onto four polyacrylamide gels (20 cm × 20 cm) containing 8 M of urea. For the gels TBE buffer (90 mM Tris, 90 mM boric acid, 2 mM EDTA) is used.

3. The RNA is detected using UV shadowing, cut out and sliced into small pieces (about 0.5 cm × 0.5 cm).

4. To elute the RNA the gel pieces are placed into 40 ml of 300 mM NaAc (pH 6.5) over night at 4 °C. The RNA is filtered (pore size 0.45 μm) to remove remaining gel residues and separated into four different 50 ml reaction tubes.

5. The RNA is precipitated by adding ethanol to a final concentration of 35 ml into each reaction tube (about 27 ml).

6. The samples are incubated at –20 °C for 1 h and afterwards the RNA is pelleted by centrifugation (1 h, 10,000 × *g*, 4 °C).

7. The supernatant is removed, each pellet is washed with 10 ml of 70 % (v/v) ethanol and the samples are again centrifuged (15 min, 10,000 × *g*, 4 °C).

8. The ethanol is removed completely and the samples are dried at room temperature for 5 min (*see* **Note 14**).

9. It is recommended to dissolve the RNA in RNase-free water but if necessary the RNA can also be dissolved in buffer.

10. After the purification the RNA has to be quality checked by gel electrophoresis.

3.7 Sample Preparation

1. Before using an aptamer, it has to be refolded properly to its folding protocol, which is individual for each aptamer.

2. To adjust the buffer to its final concentration a higher concentrated buffer is prepared. In most cases a 10× measurement buffer is convenient (*see* Subheading 2.3). Finally, the buffer is diluted by adding the RNA in the desired concentration and RNase-free water to the final volume. A good buffer to start with is 20 mM Na-cacodylate pH 6.8, 10 mM $MgCl_2$, and 100 mM NaCl [7, 8]. As the salt concentrations significantly influence the binding constant, the buffer conditions might have to be adjusted (*see* **Note 3**).

3. If no studies on the binding behavior of the aptamer exist, a good start is to titrate 100 µM of the ligand into 10 µM of the RNA. For the iTC200 instrument, 350 µl of sample volume with 10 µM of RNA are needed. The titrant is needed in a volume of 80 µl.

4. It is very important that the RNA and the ligand have the same salt concentration because otherwise heat changes, which are due to dilution effects and not due to ligand binding, can be observed. Some ligands need to be dissolved in dimethylsulfoxid (DMSO) due to higher stability. In this case the DMSO concentration needs to be considered and should not exceed 2 % (v/v).

3.8 ITC Measurement

1. In general the ITC instrument should be placed in a room with constant temperature. The instrument should be prevented from vibrations during a run. Both temperature changes and vibrations affect the instrument, which leads to inaccurate measurements.

2. Before the ITC instrument can be used for RNA samples, it should be cleaned thoroughly. Some instruments, like the iTC200 instrument, have a special cleaning program with a detergent (provided from the manufacturer) which should be used. Afterwards it is important to rinse the sample cell a couple of times with RNase-free water.

3. Before the measurement can be started, the sample has to be loaded into the sample cell. This is a very critical step because air bubbles must not be placed insight the sample cell. Remaining air bubbles completely prevent data analysis.

Table 2
Experimental parameters for the MicroCal iTC200

Total number of injections	16
Cell temperature	25 °C
Reference power	6 μcal/s
Initial delay	180 s
Syringe concentration	Concentration of the ligand in the syringe
Cell concentration	Concentration of RNA in the sample cell
Stirring speed	1000 rpm

Table 3
Injection parameters for the MicroCal iTC200

Injection volume	2.49 μl[a]
Spacing	180 s
Filter period	5 s

[a]The first injection differs from all other injections because of a volumetric error. The error occurs because of a backlash in the motorized screw used to drive the syringe [9]. As a consequence only 0.2 μl of titrant is injected in the first injection

4. Prior to each run, the sample cell should be cleaned with water and a second time with buffer to remove any residue from the previous experiment. To load the RNA sample into the cell it is very convenient to use a Hamilton pipet. For the above mentioned reasons, check the pipet for air bubbles before loading.

5. Furthermore, ensure that the sample cell is filled to the top. If the instrument has not been used for a while you might also check if the reference cell is fully filled.

6. The titrant needs to be loaded into the syringe. In most cases, the instrument is programmed to do that automatically.

7. The experimental parameters as well as the injection parameters depend on the instrument and on the aptamer. For the instrument MicroCal iTC200, one set of parameters is listed in Tables 2 and 3. The run itself is carried out by the instrument automatically (*see* **Note 15**).

3.9 Experimental Considerations

1. Only the ΔH values can be measured directly by ITC. By fitting a suitable model to the ITC isotherm, the association constant (K_a) can be calculated. As a consequence, it is important to receive maximum data points for the fitting process, as the shape of the binding isotherm dictates the accuracy of K_a.

2. The shape of the isotherm depends on K_a and the concentration of the interacting components in the sample cell.

3. The product of the stoichiometry of the reaction (n), the binding constant K_a and the concentration of aptamer in [M] provides a value which is known as the Brandt's "c" value [3].

$$c = n \times K_a \times aptamer\,conc.[M]$$

This value should be between 10 and 500 and is adjusted by varying the concentration of the aptamer in the sample cell.

4. As a control experiment, it is important to titrate the titrant into buffer without the aptamer. This experiment shows the dilution effects caused by each injection. This heat effects caused by dilution of the titrant are subtracted from the experimental data subsequently.

5. As a second control experiment, buffer has to be titrated into the aptamer in order to see heat effects, which are due to dilution of the aptamer. Normally, these effects are very small and therefore negligible.

3.10 Data Analysis

1. The thermodynamic parameters can be calculated using the software provided from the manufacturer, which is either Origin or the software provided by TA Instruments.

2. The software calculates the area below each peak. A baseline is adjusted automatically. It is important to check whether the baseline was adjusted correctly (if not it must be adjusted manually).

3. The calculated heat effects from the control experiment are subtracted from the data to get only the heat effects from the interaction between the ligand and the aptamer. Normally, this is also done by the provided software.

4. The first injection is never reliable [9]; it should be removed from the data so that it is not included into the fit. This results in a more accurate calculation of K_a.

5. A binding model is fitted to the integrals. These integrals result from the calculated area of each peak and as the concentrations of ligand after each injection are known, the data can be plotted as molar changes in enthalpy against molar ratio (Fig. 1b).

6. The fit itself is managed automatically by the software but the right binding model has to be chosen. In most cases, one ligand binds one aptamer; therefore, the one site binding model should be used. Nevertheless, there are also other binding models available which should be taken into account.

3.11 Analyzing Aptamer Mutants with ITC

In order to understand the interaction between aptamer and ligand in detail, the binding behavior of aptamer mutants can be analyzed. ITC has the advantage that both the binding enthalpy and the entropy of the binding partners can be determined; thus, the effect of the mutation on binding behavior can be identified. Nevertheless, a change in ΔG can either be due to a change in ΔH (e.g., different amount of hydrogen bonds) or due to a change in the entropy (ΔS). Such mutant analysis was done for a tetracycline binding aptamer [10]. Here, the mutants A13U and A9G both have the same ΔG values but differ in ΔH and ΔS. For A13U this is due to a reduced binding enthalpy. A9G on the other hand has a more unfavorable entropic term than the wild type aptamer. The crystal structure of the aptamer–ligand complex, which was later solved, verified these observations [11]: Whereas for A13U a direct contact between the RNA and the ligand is missing, A9 is important for the formation of the binding pocket; hence, A9U destabilizes the ground state of the aptamer.

4   Notes

1. It is important to use high copy plasmids as high amounts of plasmid are needed for large scale in vitro transcription.

2. For the complete experiment, it is important to use RNase-free water. Buffers should be autoclaved for 30 min. Gloves should be used at any time.

3. In principle, different buffers can be used. So if any further experiments are planned with the RNA, e.g., NMR spectroscopy or the analysis of UV melting profiles, it is convenient to use the same buffer for all experiments. This allows an improved comparability of the experiments. If possible, Tris buffers should be avoided as they have a strongly temperature dependent pK_a and a large heat of ionization [2]. Spermidine can be added to a final concentration of 1 mM to shield unspecific binding sites [8].

4. For ITC experiments, defined 3′-ends are not required. Nevertheless, there are many applications such as NMR spectroscopy or X-ray crystallography where it is required to have a defined RNA sequences. To be able to compare all experiments, it is necessary to use the same RNA; in this case, we recommend using a ribozyme for defined 3′-ends.

   If you do not want to use a ribozyme because defined 3′-ends are not required, you might want to use the restriction sites *Hin*dIII and *Sma*I to clone your fragment, containing the T7 promoter, into the pSP64 vector. *Sma*I has the advantage that it leaves blunt ends after digestion, which reduces the possibility that unspecific nucleotides are added.

5. Two Gs are standard and show good transcription efficiency. One G is also working but the transcription efficiency decreases

significantly. Three Gs on the other hand increase the transcription efficiency even further.

6. To generate defined 3′-ends, the HDV ribozyme is used. Before starting, it is important to check that the HDV and your RNA of choice do not have the same size. In this case it is not possible to use the HDV as the RNA and the HDV ribozyme cannot be separated by electrophoresis. In this case, the hammerhead ribozyme can be used as it is 20 bp shorter than the HDV ribozyme.

7. It is also possible to use PCR products or annealed oligonucleotides as a template for the in vitro transcription but the transcription is less efficient. If a PCR product or annealed oligonucleotides are used, the sequence 5′-CAAG-3′ should be placed downstream of the T7 promotor to increase the transcription efficiency.

   For the large scale transcription described in this article, about 1–2 mg of template are needed. The easiest way to receive such high amounts of template is to use plasmid DNA, so we recommend using plasmids.

8. For the preparation of high amounts of plasmid, the Qiagen Maxi- or Giga-preparation kit is recommended

9. We recommend digesting 4–6 mg of plasmid overnight, with 5 times less enzyme than would be needed to digest the amount of DNA in 1 h. The amount of restriction enzyme should not exceed 10 % (v/v). This is due to the glycerol in which restriction enzymes are stored. If the amount of glycerol is too high the restriction enzyme is inactive.

10. 1 volume of phenol–chloroform–isoamyl alcohol (25:24:1) is added to the digested plasmid, the sample is transferred into a MaXtract tube (Qiagen) and centrifuged. The aqueous phase is transferred into a new MaXtract (Qiagen) tube, 1 volume of chloroform–isoamyl alcohol (24:1) is added and the sample is centrifuged again. The supernatant is transferred into a new reaction tube, the plasmid DNA is precipitated by adding isopropyl alcohol and 0.1 volume of NaAc (3 M, pH 6.5). Afterwards it is incubated at −20 °C for 30 min and pelleted by centrifugation. The pellet is washed with 70 % (v/v) of ethanol, centrifuged, and after drying the pellet at 37 °C it is resolved in a suited amount of water (about 2 ml). 200 ng of plasmid DNA is quality checked on a 1 % (w/v) agarose gel.

11. A successful transcription is very often accompanied by white clouds of precipitated pyrophosphate. If there is no precipitation visible, t this does not imply that the transcription did not work. Regardless, you can still carry on with the purification.

    The success of a transcription has to be analyzed by polyacrylamide gel electrophoresis.

12. The magnesium concentration can be varied from 15 to 30 mM to optimize the transcription.

13. Before the transcription is loaded onto the gel, its salt concentration must be reduced by ethanol precipitation.

14. If the RNA is dried for too long, it is impossible to resolve it. Drying the RNA for 5 minutes at room temperature is sufficient, if the supernatant is carefully removed.

15. Total number of injections: the accuracy of the syringe dictates the number of injections because by doing too many injections the minimal injection volume is reached. Cell temperature: The K_a strongly depends on the temperature, so the temperature must be adjusted to the experimental needs. Reference Power: The reference power depends on the reaction. For strongly exothermic reactions, it should be adjusted to a value around 5 μcal/s. For very endothermic reactions, it should be adjusted to a value around 0.5 μcal/s. Initial Delay: This parameter is necessary to establish a baseline before the first injection and is specific of each experiment. Syringe and Cell concentration: These parameters depend on the K_a of the aptamer (*see* also Subheading 3.5). Stirring Speed: If the stirring speed is too low, mixing takes too long and as a consequence the spacing time has to be adjusted. However, the solution in the sample cell might contain suspended particles; then, very fast stirring is required.

Acknowledgement

This work was supported by the Deutsche Forschungsgemeinschaft (SFB902 A2) and EU FP7-KBBE-2013-7 no. 613745, Promys. We thank Katharina Keim for critical reading the manuscript.

References

1. Groher F, Suess B (2014) Synthetic riboswitches—a tool comes of age. Biochim Biophys Acta 1839:964–973

2. Ilgu M, Wang T, Lamm MH, Nilsen-Hamilton M (2013) Investigating the malleability of RNA aptamers. Methods 63:178–187

3. Wiseman T, Williston S, Brandts JF, Lin LN (1989) Rapid measurement of binding constants and heats of binding using a new titration calorimeter. Anal Biochem 179:131–137

4. Davanloo P, Rosenberg AH, Dunn JJ, Studier FW (1984) Cloning and expression of the gene for bacteriophage T7 RNA polymerase. Proc Natl Acad Sci U S A 81:2035–2039

5. Doherty EA, Doudna JA (2001) Ribozyme structures and mechanisms. Annu Rev Biophys Biomol Struct 30:457–475

6. Walker SC, Avis JM, Conn GL (2003) General plasmids for producing RNA in vitro transcripts with homogeneous ends. Nucleic Acids Res 31, e82

7. Reuss AJ, Vogel M, Weigand JE, Suess B, Wachtveitl J (2014) Tetracycline determines the conformation of its aptamer at physiological magnesium concentrations. Biophys J 107:2962–2971

8. Weigand JE, Schmidtke SR, Will TJ, Duchardt-Ferner E, Hammann C, Wohnert J, Suess B

(2011) Mechanistic insights into an engineered riboswitch: a switching element which confers riboswitch activity. Nucleic Acids Res 39:3363–3372

9. Mizoue LS, Tellinghuisen J (2004) The role of backlash in the "first injection anomaly" in isothermal titration calorimetry. Anal Biochem 326:125–127

10. Muller M, Weigand JE, Weichenrieder O, Suess B (2006) Thermodynamic characterization of an engineered tetracycline-binding riboswitch. Nucleic Acids Res 34:2607–2617

11. Xiao H, Edwards TE, Ferre-D'Amare AR (2008) Structural basis for specific, high-affinity tetracycline binding by an in vitro evolved aptamer and artificial riboswitch. Chem Biol 15:1125–1137

Chapter 10

Applications of Aptamers in Flow and Imaging Cytometry

Isis C. Nascimento, Arthur A. Nery, Vinícius Bassaneze,
José E. Krieger, and Henning Ulrich

Abstract

Aptamers compete with antibodies in many applications, in which high-affinity and specificity ligands are needed. In this regard, fluorescence-tagged aptamers have gained applications in flow and imaging cytometry for detecting cells expressing distinct antigens. Here we present prospective methods, as a starting point, for using these high-affinity ligands for cytometry applications.

Key words Aptamers, Fluorescence-labeling, Flow cytometry, Imaging cytometry, Automation

1 Introduction

Aptamers comprise a class of DNA or RNA molecules that specifically bind to their targets, rendering multiple applications. Due to their high specificity, aptamers are used as recognition molecules for virtually any target molecule class—ranging from small molecules (e.g. ampicillin or dopamine [1]) to proteins localized on a cell- or virus surface, respectively [2], or intracellularly in the cytoplasma or nucleus [3] as well as entire cell surfaces or organisms [4]. Aptamers with high-affinity have promising research, industrial, diagnostic and clinical applications. Although primary applications of aptamers aim at inhibition of target protein function [5], fluorescently labeled aptamers are more often used as tool for the identification and quantification of expression levels in western blotting, quantum dots [6], biosensors [7], imaging in fluorescence microscopy [2] and flow cytometry (Table 1).

The latter approach offers the advantage of obtaining several parameters measured simultaneously (3–16 parameters, including side scatter and forward scatter) in a cell population maintained in suspension. Furthermore, measurement precision is high in a short period of time, making flow cytometry the method of choice for experiments involving enumeration of mixed cell populations. In addition, aptamers have been developed for imaging purposes.

Günter Mayer (ed.), *Nucleic Acid Aptamers: Selection, Characterization, and Application*, Methods in Molecular Biology, vol. 1380, DOI 10.1007/978-1-4939-3197-2_10, © Springer Science+Business Media New York 2016

Table 1
Example of applications for fluorescence-labeled aptamers

Related structure	Application	Target gene	Target cell type	Fluorescent probe	References
Cell membrane	Flow cytometry and fluorescence microscopy	CD44	Cancer stem cell lines	FITC	[2]
Cell membrane	Flow cytometry and fluorescent microscopy	CD8	Lymphocytes	Cy5	[8]
Cell membrane	Flow cytometry	c-kit	BJAB lymphoblastoma cell line	PE and Alexa 647	[9]
Nuclear	Flow cytometry and fluorescent microscopy	Nucleolin	HeLa	PPIX	[3]
Cytosol	Flow cytometry	Intracellular interferon-γ	Lymphocytes	FITC	[10]
Whole bacteria	Flow cytometry	Several	*Staphylococcus aureus*	5′ABFL	[11]

FITC fluorescein isothiocyanate, *Cy5* cyanine 5, *PE* phycoerythrin, *PPIX* protoporphyrin IX, *5′ ABFL* 5′azobenzylflavin

Several features of aptamers, such as their small size, conformational flexibility and easiness of modifications, turn aptamers into excellent tools for flow cytometry and cellular imaging. Aptamers can either be covalently bound to fluorophores or nanoparticles with optical properties or non-covalently labeled by using biotin–streptavidin chemistry.

More recently, with the advancement of image processing and artificial intelligence techniques, the nominated "imaging cytometry" experiment has been developed. It consists of an automatic machine to capture high-resolution images coupled to a computer with a high-end image processing software. It captures almost the same parameters as flow cytometer with the addition of two- or three-dimensional images. Image segmentation permits subcellular identification and morphology measurements (nucleus quantities and size, cytosol area, cell perimeter, and several others). This is particularly interesting for confirm the subcellular localization of a particular target. The experiment can be performed with cells attached to a surface (rich in details) or in suspension (quicker). The experiment can be with alive (for continuous monitoring) or fixed cells (easier). With live cells, it offers the advantage of time-lapse for migration tracking experiments or intracellular changes detection of particular protein expression patterns without disrupting cell microenvironment. When information in high throughput shall be acquired, it is named "high content analysis" or "high content screening" as used for the InCell® equipment (GE LifeSciences), while "histo-cytometry" refers to this experiment in cells in an in situ context [12]. Some equipments only work with cells in suspension (Tali® Image-Based Cytometer, Life Technologies). For instance, image cytometry has been used for imaging of Ets1 expression by metastatic cells using Texas Red-labeled RNA aptamers[13].

The coupling of both aptamers and cytometry techniques produce several outcomes such as: (1) Population analysis to define the percentage of cells stained by fluorescently labeled aptamers. (2) Characterization of aptamer–target cell affinity by determination of the apparent dissociation constant (K_d of aptamers). (3) Identification of the localization of the target for a specific aptamerby imaging cytometry; (4) Determination of the cellular subtype, to which the aptamer binds by, i.e., using antibodies specific for Clusters of Differentiations (CD).

Since the experimental conditions for the use of aptamers in flow cytometry and regular fluorescence imaging experiments are essentially the same, we can conclude that both techniques work with the same methodology. Thus, here we offer a prospective method for flow cytometry that might work for imaging cytometry experiments. Bottlenecks are discussed together with troubleshooting notes.

2   Materials

2.1   Reagents (See Note 1)

1. DPBS (Dulbecco's phosphate buffer saline): 8 mM sodium phosphate, 2 mM potassium phosphate, 140 mM sodium chloride, 10 mM potassium chloride.

2. 2 mM EDTA solution diluted in DPBS.

3. Cell culture medium (according to the culture or tissue to be analyzed by aptamers).

4. Fixative solution (2×) for intracellular staining: DPBS plus specific fixative agents depending on tissue and cellular type (*see* **Note 2**).

5. Permeabilization Buffer (for intracellular staining): 0.1 % Triton X-100 in DPBS.

6. 5× Binding buffer: 25 mM HEPES, 5.3 mM KCl, 1.8 mM $CaCl_2$, 1.2 mM $MgCl_2$, 145 mM NaCl, 0.05 mg/ml yeast t-RNA, pH 7.4 (*see* **Note 3**).

7. 4.5 % (w/v) glucose solution in 1× binding buffer (*see* **Note 3**).

8. 5 % BSA solution in 1× binding buffer (*see* **Note 3**).

9. Fluorescent-labeled aptamer (*see* **Note 4**).

2.2 Others

1. 50 µm cell strainer.
2. Thermoblock.
3. Flow cytometer.
4. Centrifuge.

3 Methods

3.1 Preparation of Aptamers for Cell Labeling

1. Prepare a solution of 1.6 µM aptamers (*see* **Note 5**) in 1× binding buffer.

   (a) Example: 200 pmols of aptamer (having a 18.6 pmol/µl stock solution, we would use 10.8 µl), 40 µl of 5× binding buffer, 69.2 µl of H_2O, to a total volume of 120 µl.

2. Incubate for 10 min at 95 °C.

3. Immediately place the tube on ice for 10 min.

4. Add glucose and BSA solution to 1× at the final solution.

   (a) Example: add 40 µl of glucose and 40 µl of BSA solutions, reaching a 200 µl solution of 1 µM of aptamers in a 1× binding buffer with 0.9 % of glucose and 1 % BSA (*see* **Note 6**).

5. Incubate aptamers for 20 min at room temperature.

3.2 Preparation of Living Cells for Labeling by Aptamers

1. Remove medium. Rinse plate once with DPBS. Withdraw DPBS.

2. Add 5 ml of 2 mM EDTA solution to plate (*see* **Note 7**).

3. Mix gently and incubate at 37 °C.

4. After 5 min of incubation, mix cells gently and check for non attached cells; continue until majority of them is detached.

5. Add approximately 5 ml of culture media containing serum to dilute EDTA.

6. Remove the medium with cells and centrifuge for 5 min at $300 \times g$.

7. Remove supernatant and resuspend pellet in culture medium containing serum.

8. Gently agitate the solution for 30 min (*see* **Note 8**).

9. Centrifuge the cells, remove the medium, and resuspend them in binding buffer.

10. Pass cell suspension through a 50 µm cell strainer to eliminate cell aggregates.

11. Count the cells and separate at least 2×10^5 cells/tube.

12. Centrifuge for 5 min at $300 \times g$ for pellet formation and discard the supernatant.

13. Follow aptamer incubation procedure.

3.3 Preparation of Fixed and Permeabilized Cells

Fixation and permeabilization are required for the detection of intracellular targets. However, these procedures may increase background fluorescence and change light scatter profiles. Therefore, the same procedure must be followed for all samples, which will be compared including control samples.

1. Prepare the cells as described above for living cells. Add sufficient volume of fixative into the tube containing the cells to obtain a 1× solution.

2. Incubate at room temperature for 10 min.

3. Centrifuge for 5 min at $300 \times g$ and discard the supernatant.

4. Wash the pellet twice using 1× binding buffer + 1 % BSA.

5. Resuspend pellet in 500 µl of permeabilization buffer at room temperature and incubate for 15 min.

6. Centrifuge for 5 min at $300 \times g$ and discard the supernatant.

7. Wash the pellet twice using 1× binding buffer + 1 % BSA.

8. Centrifuge for 5 min at $300 \times g$ and discard the supernatant.

9. Follow aptamerincubation procedure and correspondent washing steps.

3.4 Incubation of Cells with Aptamers

1. In the tube, gently mix together the cell pellet (2×10^5 cells/ tube) and aptamer solution (1 µM final concentration to a final volume of 200 µl).

2. Incubate mixture at room temperature gently shaking for 30 min.

3. Centrifuge for 5 min at $300 \times g$ and remove the supernatant.

4. Add 200 µl of binding buffer to cells.

5. Centrifuge for 5 min at $300 \times g$ and remove the supernatant.

6. Resuspend pellet in 500 µl binding buffer.

7. Use this sample for flow cytometry analysis.

3.5 Cytometry Controls

Controls must ensure to comprise:

1. Control 1: cells without any aptamer or antibody for autofluorescence determination.

2. Control 2 (or more): cells marked with only aptamers or any other marker of interest (antibodies for example) for channel leak, proper compensations and gating prior to experiments (*see* **Note 9**).

3.6 Determination of the Dissociation Constant (K_d) of Aptamer–Cell Binding with Flow Cytometry

The assessments for the target antigen-binding equilibrium dissociation constant (K_d) of aptamers in cytometry must be performed with single aptamers rather than with populations or pools of different molecules. Therefore, prior to cytometry assays, sequencing, clustering, and structural analysis of the selected molecules through the SELEX technique must be already performed.

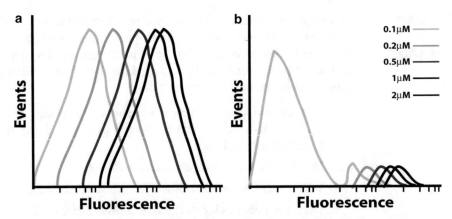

Fig. 1 Profile of fluorescence gain according to increment of aptamer concentration. (**a**) Increase of fluorescence for a population completely marked by fluorescence-tagged aptamers. (**b**) Increase of fluorescence for a subpopulation marked by a fluorescence-tagged aptamers

1. In order to assess the K_d of aptamer–cell binding, the mean fluorescence intensity of the stained population labeled by several concentrations of target aptamer is determined [9]. The equation

$$Y = B_{max} \cdot X / (K_d + X) \qquad (1)$$

   is used to calculate the K_d of aptamer binding, where B_{max} is the maximum percentage of fluorescence, Y is the mean percentage fluorescence, and X is the molar concentration of the aptamer.

2. When the K_d of aptamer binding to the entire population shall be determined, the mean fluorescence intensity of the population must be considered (Fig. 1a). For measuring K_d values of aptamers that bind just to cells within a subpopulation, these need to be gated for proper quantification (Fig. 1b).

3.7 Analysis of Aptamer-Target Cell Binding by Imaging Cytometry

1. Preparation of samples for imaging cytometry follows the same rationale described for flow cytometry. The difference for imaging cytometry is that the samples instead in solution analyzed in the flow cytometer, while they are plated on slides for experiments with an imaging cytometer.

2. Both cytometry techniques are complementary, whereas one has a power of measuring larger cell populations (flow cytometry), while imaging cytometry conjugates the visual analysis for better tracking of aptamer localizations with antigen-immunostaining by antibodies.

3.8 Characterization of the Cellular Subtype Recognized by Aptamers Using Antibodies Directed against Clusters of Differentiations (CD)

Both cytometry techniques (flow and imaging cytometry) can be combined for better understanding, to which cell type fluorescence-labelled aptamers bind. Specific surface antigens are expressed depending on the cell type.

By using flow cytometry co-expression of aptamer targets and CD antigens can be verified, while imaging cytometry reveals further details on the morphology of the cells. Moreover, imaging cytometry possesses advantages over flow cytometry, when adherent cells shall be analyzed. The dissociation process using trypsin or EDTA damages cell surfaces and may affect cell viability and antigen expression patterns.

4 Notes

1. Prepare all solutions using ultrapure water (prepared by purifying deionized water to attain a resistivity of 18 MΩ cm at 25 °C) and analytical grade reagents. Prepare and store all solutions at room temperature (unless indicated otherwise) and all reagents for molecular biology at 4 °C. Diligently follow all waste disposal regulations when disposing waste materials.

2. Several fixative agents can be used, such as paraformaldehyde, methanol and acetone, ranging from 1 to 4 % depending on the tissue or cell type. We suggest prior standardization to optimize the fixation process for each cell type to be analyzed by flow or image cytometry.

3. This buffer should be filtered and stored at 4 °C

4. The selection of the most adequate fluorochrome combination depends on the specific optical configuration of the flow cytometer.

5. The final concentration of aptamers used for cytometry depends on the aptamer affinity for the target and experimental design according to the hypothesis to be tested.

6. Do not add the glucose or BSA prior to heat denaturation, as they can make the solution too viscous.

7. Trypsin solution can compromise cell surface targets and alter immunostaining results. For cells with highly expressed target antigens, trypsin is a valid option to speed up the removal of cells.

8. This step allows the cells to assume a more homogeneous shape, facilitating gating for side and forward scatter parameters in flow cytometry.

9. We also recommend the use of the Viability Dyes to help eliminating dead cells during cytometry analysis.

Acknowledgments

This work was supported by Thematic Grants from Fundação de Amparo à Pesquisa do Estado de São Paulo (FAPESP, project No. 2012/50880-4 to H.U. and No. 2013/17368-0 to J.E.K.) and by grants from Conselho Nacional de Desenvolvimento Científico e Tecnológico (CNPq, project No. 486294/2012- 9 and 467465/2014-2 to H.U.), Brazil. V.B. acknowlegdes postdoctoral fellowship support by FAPESP (project No. 2011/19678-1).

References

1. Walsh R, DeRosa MC (2009) Retention of function in the DNA homolog of the RNA dopamine aptamer. Biochem Biophys Res Commun 388:732–735

2. Ababneh N, Alshaer W, Allozi O, Mahafzah A, El-Khateeb M, Hillaireau H, Noiray M, Fattal E, Ismail S (2013) In vitro selection of modified RNA aptamers against CD44 cancer stem cell marker. Nucleic Acid Ther 23:401–407

3. Ai J, Li T, Li B, Xu Y, Li D, Liu Z, Wang E (2012) In situ labeling and imaging of cellular protein via a bi-functional anticancer aptamer and its fluorescent ligand. Anal Chim Acta 741:93–99

4. Shin S, Kim IH, Kang W, Yang JK, Hah SS (2010) An alternative to Western blot analysis using RNA aptamer-functionalized quantum dots. Bioorg Med Chem Lett 20:3322–3325

5. Ulrich H, Wrenger C (2009) Disease-specific biomarker discovery by aptamers. Cytometry A 75:727–733

6. Faria M, Ulrich H (2008) Sugar boost: when ribose modifications improve oligonucleotide performance. Curr Opin Mol Ther 10: 168–175

7. Wang RE, Zhang Y, Cai J, Cai W, Gao T (2011) Aptamer-based fluorescent biosensors. Curr Med Chem 18:4175–4184

8. Zhang P, Zhao N, Zeng Z, Chang CC, Zu Y (2010) Combination of an aptamer probe to CD4 and antibodies for multicolored cell phenotyping. Am J Clin Pathol 134: 586–593

9. Meyer S, Maufort JP, Nie J, Stewart R, McIntosh BE, Conti LR, Ahmad KM, Soh HT, Thomson JA (2013) Development of an efficient targeted cell-SELEX procedure for DNA aptamer reagents. PLoS One 8, e71798

10. Cao B, Hu Y, Duan J, Ma J, Xu D, Yang XD (2014) Selection of a novel DNA aptamer for assay of intracellular interferon-gamma. PLoS One 9, e98214

11. Baumstummler A, Lehmann D, Janjic N, Ochsner UA (2014) Specific capture and detection of Staphylococcus aureus with high-affinity modified aptamers to cell surface components. Lett Appl Microbiol 59: 422–431

12. Gerner MY, Kastenmuller W, Ifrim I, Kabat J, Germain RN (2012) Histo-cytometry: a method for highly multiplex quantitative tissue imaging analysis applied to dendritic cell subset microanatomy in lymph nodes. Immunity 37: 364–376

13. Kaur J, Tikoo K (2015) Ets1 identified as a novel molecular target of RNA aptamer selected against metastatic cells for targeted delivery of nano-formulation. Oncogene. doi: 10.1038/onc.2014.447. [in press]

Chapter 11

In Vitro and In Vivo Imaging of Fluorescent Aptamers

Ioanna Théodorou, Nam Nguyen Quang, Karine Gombert, Benoit Thézé, Benoit Lelandais, and Frédéric Ducongé

Abstract

Fluorescence imaging techniques could be used in different ways to study the interaction of aptamers with biological systems from cell culture to animal models. Here, we present the methods developed in our laboratory for fluorescently labeled aptamers, study their internalization inside living cells using time-lapse microscopy, and monitor their biodistribution in mice bearing subcutaneous xenograft tumors using planar fluorescence imaging and fluorescence diffuse optical tomography (fDOT).

Key words Aptamer, Oligonucleotide, Fluorescence, Time-lapse microscopy, Fluorescence diffuse optical tomography

1 Introduction

Nucleic acid aptamers could be developed as tools for research, diagnosis, or therapy [1–3]. Whether it is used for itself or to address contrast agents for imaging or therapeutic payloads, it is often necessary to study the interaction of an aptamer with biological tissues. For such studies, optical imaging and particularly fluorescence imaging could be useful to visualize, characterize, and quantify biological processes at the molecular and cellular levels. For instance, fluorescence imaging can be used to study the binding of an aptamer to its target, its internalization by a specific cell, or its biodistribution in a preclinical disease model [4–14]. Studies could be performed in vitro on cell culture or ex vivo from biopsy using fluorescence microscopy. Additionally, fluorescence imaging is now increasingly used directly in vivo in small animals [14–16]. First, "planar imaging" has been established for qualitative or semiquantitative two-dimensional imaging on the surface of animals. Such imaging technique is simple and involves a wide-field

Günter Mayer (ed.), *Nucleic Acid Aptamers: Selection, Characterization, and Application*, Methods in Molecular Biology, vol. 1380, DOI 10.1007/978-1-4939-3197-2_11, © Springer Science+Business Media New York 2016

illumination of the animal for the fluorophore excitation while a highly sensitive detector such as a CCD camera is used for capturing the fluorescence signals using appropriate filters. However, light photons are subject to high amounts of absorption and scattering when traveling through tissues. These phenomena prevent proper visualization and quantification of the fluorescence signal when it originates from a depth greater than a few millimeters. To solve this problem, fluorescence diffuse optical tomography (fDOT) technique, also known as fluorescence molecular tomography (FMT), has been developed. It involves instruments that use a transillumination mode for excitation generally by lasers, and sophisticated reconstruction algorithms to reassign the original fluorescence emission signal in three dimensions (3D) [16]. Although fDOT is still restricted to perform imaging inside tissue of a few centimeters in depth, it is perfectly adapted for in vivo imaging in rodents. Furthermore, we previously demonstrated that fDOT, in a comparable manner to positron emission tomography (PET), can quantify the biodistribution of a fluorescent oligonucleotide inside mice for concentrations ranging from 3 nM to 1 μM [15]. In contrast to nuclear imaging technologies such as PET, fluorescent imaging techniques have the advantage of being cheaper and easier to implement since it does not require the radiation protection infrastructure that is necessary for the production of isotopes and the radiochemistry. Therefore, fluorescence imaging techniques have already been used by several groups and us to study from in vitro to in vivo proofs of mechanism and proofs of efficacy of aptamers [4–12, 14].

In this chapter, we describe methods for (1) the labeling of aptamers with fluorescent dyes, (2) in vitro time-lapse microscopy on living cells, and (3) in vivo fluorescence imaging of aptamers in mice bearing subcutaneous tumors. These procedures can be used and adapted to screen series of aptamers in just a few days and are particularly adapted for aptamers that bind membrane proteins.

2   Materials

2.1   Fluorescent Labeling of Aptamers

1. ULYSIS® Nucleic Acid Labeling Kits (Life Technologies, USA).

2. 3 M Sodium acetate solution, pH 5.2.

3. Absolute ethanol.

4. (Optional) linear polyacrylamide (LPA).

5. Heat block.

6. Micro Bio-Spin® Columns with Bio-Gel® P-6 (BioRad, USA).

7. 1.5 mL tube centrifuge.

8. UV-visible spectrophotometer.

9. Vortex.

2.2 In Vitro Imaging of Aptamers Using Time-Lapse Microscopy

1. Epifluorescence videomicroscope equipped with an incubation chamber.

2. Glass-bottom dishes (MatTek/Nunc).

3. Collagen type I solution from rat tail.

4. Phosphate buffer saline (pH 7.4) with $MgCl_2$ (0.1 g/L) and $CaCl_2$ (0.133 g/L).

5. 0.22 μm filter.

6. RPMI 1640 Glutamax.

7. $MgCl_2$ (50 mM), tRNA (10 mg/mL), dextran sulfate (33.33 mg/mL, MP Biomedicals, USA), and heparin Choay (5000 UI/mL, SANOFI-AVENTIS, France).

8. Bovine serum albumin (2 mg/mL, VWR, USA).

9. Hoechst 33342, lysotracker, mitotracker, endoplasmic reticulum (ER) tracker, AF680-transferrin, and AF488-wheat germ agglutinin (Life Technologies).

2.3 Cells

MCF-7 cells (human breast adenocarcinoma) from ATCC were cultured at 37 °C with 5 % CO_2 in RPMI 1640 Glutamax medium completed with 10 % heat-inactivated fetal bovine serum and 1 % antibiotic/antifungal solution. The cell population is usually seeded at a density around 50,000 cells/cm^2 and allowed to adhere for about 48 h. Afterwards, the medium is removed but 1–2 mL is left in the flask. The cells are then carefully scraped off the flask surface, transferred in a Falcon tube, and total cell count is recorded. The cell solution is centrifuged before being diluted in PBS and stored on ice until subcutaneous injection in mice in order to induce tumor xenografts.

2.4 Subcutaneous XenograftTumor Model in Mice

1. Nude mice (NMRI-Foxn1nu/Foxn1nu).

2. 17-Beta estradiol pellets (0.72 mg each, Innovative Research of America).

3. Stainless steel reusable precision trochar (Innovative Research of America).

4. 1 mL sterile syringe with 29 G needle (TERUMO).

5. Matrigel basement membrane matrix (Corning).

6. Phosphate buffer saline (pH 7.4) with $MgCl_2$ (0.1 g/L) and $CaCl_2$ (0.133 g/L).

7. Anesthesia system equipped with anesthesia chamber and coaxial breathing nozzle (Equipement Vétérinaire Minerve, France).

8. Isoflurane (Inhalation liquid, 100 mL, Abbott, France).

2.5 Instrument for In Vivo Fluorescence Imaging

In our laboratory, we use a fluorescent diffuse optical tomography (fDOT) imaging system (developed by CEA/LETI and Cyberstar [17]). This imaging system is composed of near-infrared LEDs and lasers for illumination, and a charge-coupled device (CCD) camera and filters for detection [17]. The CCD camera is focused at the top surface of the animal. This instrument can be used for acquiring 3-dimensional images but it can also be used for planar imaging. For 3D imaging, the algorithm used for our imaging system is presented in [18]. By specifying the required voxel size parameter, the algorithm allows retrieving the fluorescence in a stack of 2-dimensional images whose number depends on the animal thickness and whose size depends on the laser scan area. After image reconstruction, the voxels are $0.6 \times 0.6 \times 1$ mm^3 depending on reconstruction parameters.

3 Methods

3.1 Fluorescent Labeling of Aptamers

This section presents an easy and straightforward method to fluorescently label aptamers. This method can be used with DNA or RNA aptamers since it use commercially available platinum dye complexes that form a stable adduct with the N7 position of guanine and, to a lesser extent, adenine bases (*see* Fig. 1) [19]. The manufacturer's protocol is optimized to prepare DNA hybridization probes but it can be adapted for the labeling of aptamers. We present here experimental conditions for the labeling of aptamers with Alexa Fluor® 546 nm or Alexa Fluor® 680 nm (*see* **Note 1**).

3.1.1 Titration of the Aptamer Solution

1. Measure absorbance of the aptamer solution at $\lambda = 260$ nm using a UV-visible spectrophotometer.

2. The RNA concentration can be calculated from the Lambert-Beer law, which is concentration (mol/L) = absorbance (260 nm)/$(\varepsilon \lambda_{=260} \times$ path length (cm)).

For each aptamer, its extinction coefficient at 260 nm can be calculated on the basis of its sequence, applying the method published by Tataurov et al. [20].

3.1.2 RNA Aptamer Precipitation

1. Pipette the required volume of the stock solution to collect 0.5 nmol of aptamers and complete the volume to 100 μL with nuclease-free water (*see* **Note 2**).

2. Add to the aptamer solution, 10 μL of 3 M sodium acetate solution (pH 5.2), and 500 μL of ethanol 100 %.

3. Optionally add 5 μL of linear polyacrylamide, a synthetic coprecipitant, to maximize RNA precipitation and recovery.

4. Vortex for 1 min.

5. Centrifuge at $20,000 \times g$ (20 min, 4 °C) and discard the supernatant.

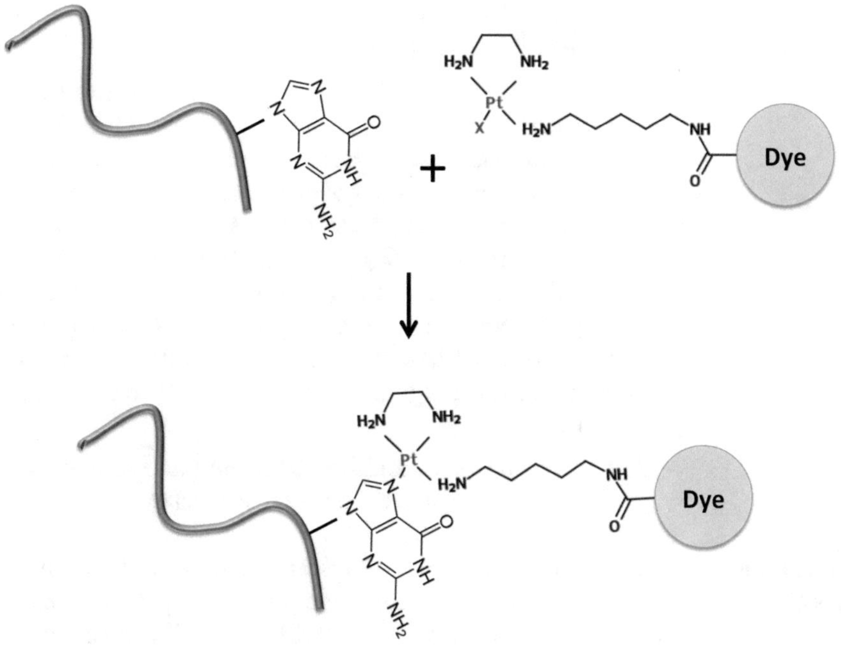

Fig. 1 Schema of the labeling method using platinum dye. A platinum group reacts with the *N*-7 position of guanine residues to provide a stable coordination complex between the oligonucleotide and the dye

3.1.3 Conjugation of the Alexa Fluor® ULS Reagent with the Aptamer

1. Resuspend dehydrated ULS labeling reagent in 50 μL of 50 % dimethylformamide (DMF) or 100 % dimethylsulfoxide (DMSO) as provided in the kit.

2. Resuspend immediately the pellet in 25 μL of labeling buffer.

3. Denature RNA secondary structures at 90 °C for 5 min and snap cool on ice. Centrifuge briefly to redeposit the sample to the bottom of the tube.

4. Add 4.5 μL of Alexa Fluor® 546 ULS labeling reagent or 0.5 μL of Alexa Fluor® 680 ULS labeling reagent.

5. Heat the reaction mixture at 90 °C for 15 min.

3.1.4 Purification of the Fluorescent Aptamer

The labeled aptamer has to be purified from the excess ULS labeling reagent using gel filtration-based spin column.

1. Vortex Micro Bio-Spin® Columns with Bio-Gel® P-6 and eliminate bubbles. Let saline-sodium citrate (SSC) buffer drain by gravity during a few seconds.

2. Centrifuge the column at $1,000 \times g$ (4 min, 4 °C) and collect the SSC buffer, later used as blank reference for the titration of the final product by spectrophotometry.

3. Load the reagent mix on the Bio-Spin columns (20–75 μL per column).

4. Centrifuge at $1,000 \times g$ (4 min, 4 °C) and collect the eluate containing the purified labeled aptamer.

3.1.5 Calculation of the Labeling Yield	1. Acquire the absorbance spectrum of the labeled aptamer in the range of $\lambda = 240$–800 nm.

2. As previously described, the aptamer concentration should be calculated from the measured absorbance at $\lambda = 260$ nm (A_{260}). The fluorophores present an absorption pic (A_{dye}) at their maximum excitation wavelength, and the dye concentration can be determined from the measured absorbance value at this wavelength, using $\varepsilon\lambda_{=546} = 104{,}000$ cm.M^{-1} and $\varepsilon\lambda_{=680} = 164{,}000$ cm.M^{-1} for Alexa Fluor® 546 and Alexa Fluor® 680 nm, respectively. The ratio C_{dye} (μmol/L)/$C_{aptamer}$ (μmol/L) gives the degree of labeling, which is usually, in the above conditions, in the range between 0.9 to 1.7 dyes per aptamer (*see* **Note 3**).

3. Labeled probes should be aliquoted and stored frozen in the dark. Avoid repeated freeze/thaw cycles.

3.2 In Vitro Imaging of Aptamers Using Time-Lapse Microscopy

The cellular uptake of fluorescent probes can be monitored using epifluorescence time-lapse videomicroscopy. Colocalization of fluorescently labeled aptamers with intracellular organelle markers permits to give insight about the pathways involved in their internalization and about their fate inside the cells.

3.2.1 Preparation and Culture of the Cells

High-resolution fluorescence microscopy requires the use of glass-bottom dishes with high-quality #1.5-thick German glass. Those products are commercially available in the Petri dish or multi-well plate format, as well as in the Lab-Tek chamber slide format. As cellular adherence on glass is limited, surface coating with an extracellular matrix component is required. Collagen type I solution from rat tail is suitable for this as its net charge is close to zero at pH 7–8, which avoids unwanted electrostatic interactions with the aptamers. Perform the coating as described below:

1. Fill the bottom of each well with a 0.22 μm filtered 50 μg/mL collagen solution prepared with 0.02 N acetic acid in phosphate buffer saline (PBS).

2. Incubate for 1 h at room temperature (R/T).

3. Discard the collagen solution and rinse with PBS (2–3 times).

4. Let dry and store at 4 °C until use.

5. Seed the cell population at a density ranging from 10,000 to 50,000 cells/cm^2 and allow them to adhere for about 48 h according to the strain.

6. Cells can be cultured in their usual medium, but sources of autofluorescence have to be avoided during the microscopy experiment, decreasing the fetal bovine serum concentration and banishing the use of phenol red-containing media (*see* **Note 4**).

1. Set the incubation chamber to appropriate cell culture conditions (37 °C, 5 % CO_2, humid atmosphere) and let equilibrate the imaging system for at least 1 h.

2. Choose high numerical aperture (NA) objective, set the fluorescence source lamp illumination as low as possible (less than 10–20 % of the maximal power for a 120 W lamp with an ×63 objective (NA = 1.4)), minimize bright-field illumination as well (5 V, NA condenser = 0.3) and reduce the field diaphragm apertures to the field of view in order to limit cell photo-damages and probe photo-bleaching. If fluorescence quantification is not on purpose, gain can be set on the CCD camera at the cost of a loss in the sensor response linearity. Otherwise, pixel binning can be set, improving the signal-to-noise ratio and preserving the signal linearity but reducing the image resolution. Camera exposition times for each channel should be less than 500 ms, around 100 ms being better.

3. Incubate the cell monolayer with 10 nM fluorescent aptamer solution (15–30 min, 37 °C) in RPMI medium without phenol red mix with $MgCl_2$ (5 mM), tRNA (100 μg/mL), and bovine serum albumin (1 mg/mL). Moreover, the binding conditions can be optimized to reduce nonspecific signals using dextran sulfate (10 %, Sigma-Aldrich), and heparin (500 μg/mL) [21].

4. If needed, the cellular organelles can be stained adding Hoechst 33342 (5 μg/mL), lysotracker (75 nM), mitotracker (500 nM), endoplasmic reticulum (ER) tracker (1 μM), fluorescently labeled transferrin (100 μg/mL), and fluorescently labeled wheat germ agglutinin (5 μg/mL) for, respectively, nucleus, lysosome, mitochondria, endoplasmic reticulum, early endosomes, and cytoplasmic membrane labeling. If the aptamer binding time is short (less than 30 min), preincubate the cells with organelle stains to obtain a sufficient labeling intensity. At the opposite, cytoplasmic membrane staining time has always to be shorter than 10 min, as the wheat germ agglutinin tends to be internalized by endocytic pathway.

5. At the end of the probe incubation time, wash the wells for three times with phosphate buffer saline. During the acquisition period, cellular organelle markers can be added, if necessary, at a lower concentration to sustain the fluorescence staining level (*see* **Note 5**).

6. Configure the microscope settings for time-lapse experiment, periodically acquiring images for the next 24 h if required, in order to follow the aptamers inside the cellular machinery (*see* Fig. 2).

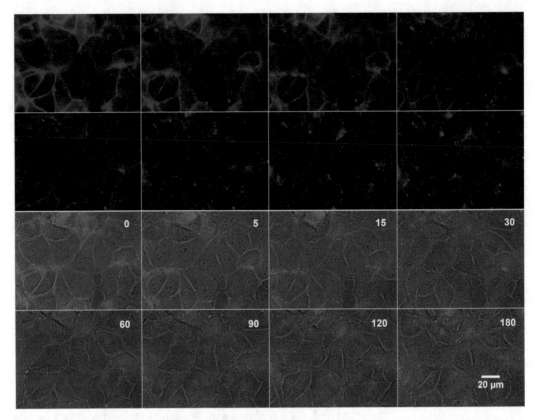

Fig. 2 Example of time-lapse imaging over 3 h of an anti-Annexin A2 aptamer on MCF7 cells. The fluorescently labeled aptamer was incubated with MCF-7 cells before washing and fluorescent time-lapse imaging. The *first two rows* display the red fluorescent channel only. The *two second ones* are a merge of the corresponding fluorescent and bright-field channels. The elapsed time is expressed in minutes and recorded on each merged image only. Most amount of the aptamer was localized at the cell surface after 30 min of incubation. Then, it progressively disappears from the cell surface and increases inside intracellular vesicles. These results demonstrate the usefulness of fluorescence microscopy to study the internalization of aptamers inside cells. Figure from the original research paper [14]

3.3 In Vivo Imaging of Aptamers

3.3.1 Subcutaneous Xenograft Tumor Model in Mice

Animal experiments are subject to approval by an animal ethics committee prior to experimentation (*see* **Note 6**). The fur of animals is not compatible with fluorescence imaging because it increases adsorption and diffusion of the light. Use either nude mice or remove the fur prior to imaging. In our laboratory, we mostly use female nude mice, with a weight of approximately 20 g. Mice are housed under standard conditions with food and water *ad libitum* and the tumor progression is daily monitored.

1. For human MCF-7 breast cancer subcutaneous tumor model, 1 week before the implantation of the MCF-7 cancer cell line, one pellet of 17-beta estradiol (0.72 mg) is subcutaneously implanted in an anesthetized mouse using a precision trochar. It is mandatory to increase the level of estrogen in nude mice for this model because MCF-7 is an estrogen-dependent cell line.

2. For implantation of cells, a sterile 29 G syringe is prepared containing 10^7 MCF-7 cells in a volume of 100 µL of PBS mixed with 100 µL of Matrigel basement membrane matrix at 4 °C.

3. Anesthetize the animal by placing them in a chamber filled with 3 % isoflurane gas and a flow of medical oxygen. Monitor their respiration, heart rate, and temperature frequently.

4. Remove the animal from the anesthesia chamber and place it on a breathing nozzle that maintains it with 1.5–2.5 % of isoflurane. In our laboratory, the nose of the animal is introduced in a coaxial breathing nozzle.

5. Carefully clean the skin between the shoulders with a medical disinfectant solution.

6. Finally, the cells are subcutaneously implanted with a sterile 29 G syringe between the shoulders of anesthetized mice and are allowed to grow for several weeks until the desired tumor size is reached (~300 mm^3 after 10 weeks for MCF-7 xenograft model).

3.3.2 Planar Near-Infrared Fluorescence Imaging

1. Prior to whole-body planar near-infrared fluorescence imaging, mice are anesthetized with 4–5 % isoflurane gas and a flow of medical oxygen. Afterwards the level of isoflurane concentration is lowered down to 2–2.5 %. During anesthesia all animals are maintained normothermic (body temperature: 36.7 ± 0.5 °C) through the use of either the anesthesia induction chamber or the heating stage of the fDOT apparatus.

2. Before injecting the fluorescent aptamer, the natural autofluorescence of the mice is recorded (*see* **Note 7**). This signal will be subtracted later on in order to obtain the accurate fluorescence signal emanating only from the injected fluorescent probes. For that carefully place the mice on their dorsal side in the fDOT apparatus making sure that they are well positioned and breathing normally through the adapted anesthesia tubing in the apparatus. Choose the appropriate excitation and emission filter sets, the exposition time and finally acquire fluorescence images with the acquisition software of the fDOT system. The acquisition of planar images is based on the excitation of fluorophores by the LEDs (emitting light between 650 and 670 nm) placed above the animal and on the reception of the fluorescent signal using the CCD camera and a band-pass filter (700–740 nm) or a long-pass filter (750 nm). Before acquisition, the user has to specify the exposition time. The higher the fluorescence is, the lower the exposition time will be. The user modifies its value until the final obtained image is unsaturated. Repeat the image acquisition by turning the mice on their ventral and left lateral sides.

3. Inject the fluorescent aptamer in the tail vein using a 29 G (insulin-type) syringe. Injected volume should not exceed 150 μL.

4. Acquire fluorescence images just after injection (1 min post-injection) of dorsal, ventral, and left lateral sides of the mice using the appropriate excitation and emission filter sets and adjusting the integration time. In our experience, a good contrast is obtained after exposition times of a few milliseconds. Repeat over various time points according to your imaging protocol (for example, acquire images at 1, 10, 30, 60, etc. minutes post-injection) (*see* **Note 8**).

3.3.3 Fluorescence Diffuse Optical Tomography

The biodistribution of aptamers can be measured using fDOT imaging. The fDOT imaging principle is based on the transillumination of a laser (690 or 740 nm) into the animal and on the sequentially reception, using the CCD camera and the filters, of the diffused excitation light and of the fluorescence signal. By moving the laser below the animal, large areas can be retrieved. A green planar laser is also used for detecting the thickness of the animal. This set of measurements (fluorescence, diffusion, and thickness) is used for reconstructing 3D fluorescence images. The protocol for fDOT imaging acquisition is as follow:

1. The mice are positioned only on their dorsal side and, using the acquisition software, a region of interest (ROI) is drawn on the tumor region (usually the ROI is a 5×5 grid that correspond to an area about 12×12 mm^2). To help the user to choose the position of the laser depending of the position of the tumor, a white planar image of the mouse is also acquired. The emission filter sets are chosen according to the emission wavelength of the fluorophore.

2. Acquire 3D images using the acquisition software. The system first acquires the transmitted (excitation) images and subsequently the emission filter set is used to record the fluorescence (emission) images.

3. The 3D scan takes a few minutes and is usually performed between planar imaging acquisitions.

4. For fDOT imaging, the reconstructions of the fluorescence signal distribution in the tumor tissue are performed by the TomoFluo3D built-in reconstruction software. Image reconstruction is an inverse problem that consists in retrieving a 3D image from a set of planar images.

3.4 In Vivo Molecular Imaging Analysis

Fluorescence image analysis is made using ImageJ [22]. It is powerful image analysis software that allows tracing regions of interest, applying thresholds, and making measurements as sum, mean or standard deviation. Furthermore, it is possible to integrate plug-ins developed in JAVA programming language. For this purpose, we developed and integrated the well-known Fuzzy C-Means clustering method [23]. It allows us to automatically determine the threshold to apply by analyzing the fluorescence distribution in the region of interest vs. in the background. This section presents how to analyze both planar and 3D images using ImageJ.

3.4.1 Planar Image Analysis

Because planar imaging does not allow us to visualize deep tissues, the following protocol is mainly used for subcutaneous tumor xenografts.

1. An ROI is drawn in the background of one random image from a sequence of time images and the mean of fluorescence in this region is measured (corresponding to F_1). Then, the intensity of each pixel for each image is subtracted from this value. Finally, for each image, the intensity of each pixel is divided by its exposition time. This first step allows us to normalize the images. Hence, if $Fi(t)$ is the fluorescence intensity of a pixel i at time t, and $T(t)$ is the exposition time used for the acquisition of the image. The normalized fluorescence for pixel i at time t, noted $F'i(t)$, is given by:
$$F_i'(t) = (F_i(t) - F_1) / T(t)$$

2. Using normalized images, an ROI is manually drawn for each time to delineate the tumor (*see* Fig. 3a), and the mean of intensity in this region ($F'_2(t)$) is subtracted by the mean of intensity in the same area before injection ($F'_2(t_0)$) which corresponds to the autofluorescence of the animal at time t_0.

3. Using normalized images as well, an ROI is manually drawn for each time point to delineate a muscular area in the leg (*see* Fig. 3a) and the same procedure previously described in **step 2** is applied. Hence, the mean of fluorescence at time t in a muscle area can be measured ($F'_3(t)$) as well as the autofluorescence of the animal in the same region at time t_0 ($F'_3(t_0)$).

4. The tumor targeting of the aptamer can be evaluated by the tumor/muscle ratio at each time point represented by the equation $(F'_2(t) - F'_2(t_0))/(F'_3(t) - F'_3(t_0))$. This ratio indicates if the aptamer has a higher uptake by the tumor compared to the muscle. However, this ratio has to be statistically compared with the ratio measured for a scramble sequence in order to conclude if the tumor uptake is specific. Indeed, oligonucleotides could sometime present high contrast in the tumor compared to muscle due to leaky vessels in the tumor (*see* Fig. 3b).

3.4.2 3D Image Analysis

1. An ROI encompassing the tumor is drawn such that the fluorescent regions that are outside the tumor area are excluded. This step can be made easier by merging 3D fluorescent imaging with anatomical imaging such as computed tomography or

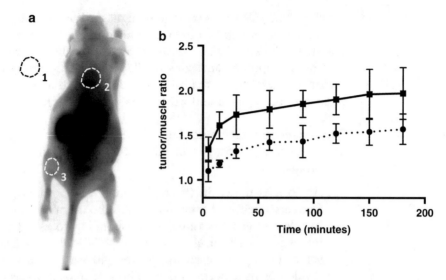

Fig. 3 Evaluation of tumor targeting by planar fluorescence imaging. (**a**) Example of fluorescence image obtained after the injection of a fluorescent aptamer in a mouse bearing a subcutaneous tumor xenograft. Region 1 depicts the chosen area for measuring the background intensity (F_1), region 2 depicts the area for measuring the fluorescence signal from the tumor ($F_2(t)$ and $F_2(t_0)$) and region 3 depicts the area for measuring the fluorescence signal from the muscle ($F_3(t)$ and $F_3(t_0)$). (**b**) Example of tumor/muscle ratio kinetic for an aptamer (*filled line*) and a nonspecific (*scramble*) oligonucleotide (*dotted line*). The fluorescent aptamer has a higher tumor targeting than the control

magnetic resonance imaging. If they are not available, it is also possible to superimpose 3D fluorescent imaging on white planar imaging (*see* Fig. 4).

2. In order to separate the fluorescent signal from the background, i.e., to segment the fluorescent region, we use either the automatic Fuzzy C-Means clustering method or a threshold fixed at 10% of the maximal recorded intensity. The threshold was empirically chosen using acquired images of capillaries filled with fluorophores.

3. The fluorescence mean concentration inside the segmented region is measured. At this last step, the measurement does not correspond to the actual fluorophore concentration in the tumor. To quantify this concentration, a calibration curve is used (*see* **Note 9**).

4   Notes

1. For in vivo imaging it is important to use dyes (such as Alexa Fluor® 680) that emit light in the near-infrared region (700–800 nm) because the absorption of photons by biological components is reduced at these wavelengths. However such dye

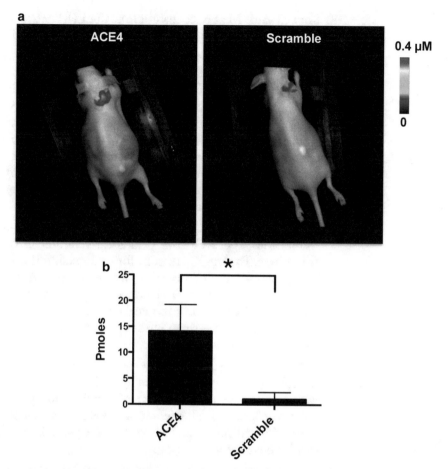

Fig. 4 Example of tumor targeting of an anti-Annexin A2 aptamer measured by fluorescence diffuse optical tomography (fDOT) imaging. (**a**) Visualization in the bird's-eye view of the 3D reconstructed fluorescent signal in an MCF-7 subcutaneous tumor (color LUT) overlaid on the white light image of the mouse for the ACE4 aptamer (*left panel*) and the control sequence (*right panel*). The acquisitions were performed 3 h post-intravenous injection. (**b**) Quantity of oligonucleotide inside tumors calculated from the 3D fluorescent signal (**a**). Error bars represent the standard deviation of triplicate. *$P < 0.05$. Figure from the original research paper [14]

could be difficult to detect by fluorescence microscopy that usually use CCD camera that are optimized to detect photons in the visible range. That is why we used Alexa Fluor® 546 for microscopy studies.

2. Classically, for live cell imaging purpose, 0.5 nmol of aptamers are labeled following this procedure. Higher amounts can be prepared but labeling buffer and ULS labeling reagent volumes have to be increased accordingly.

3. As fluorescence auto-quenching was observed when the number of fluorophores per aptamer molecule increases, the aim is to conjugate about only one dye per probe.

4. Roswell Park Memorial Institute medium (RPMI) is a bit less fluorescent than Dulbecco's modified Eagle's medium (DMEM).

5. The acquisition of z-stacks with a fast microscopy system permits to apply 3D iterative image deconvolution [24] with the ImageJ software [25]. This produces images with increased resolution, better contrast, and improved signal-to-noise ratio. Then, the JaCoP plug-in [26] is suitable for colocalization studies. Moreover, 2D-image analyses can be led using the open-source software CellProfiler [27]. The latter permits object-based segmentations, as well as feature (intensity, shape, texture) extraction and classification, opening the way to spatial and temporal distribution studies.

6. Depending on the regulations applied at your institution, you should seek approval from your local veterinarian or animal committee. Explain the reasons for undertaking the study and justify the number of animals that will be used. Anticipate all procedures in order to limit unnecessary pain to the animals. Only trained staff under the authority of a researcher should conduct animal experimentations with registered certification for animal laboratory science.

7. Usually the highest autofluorescence signal is recorded from the spleen or the intestine. If imaging is performed in a model with a tumor in the proximity of the intestine (for example colon cancer models), the autofluorescence signal should be reduced in the intestine feeding the mice with chlorophyll free diet 15 days before imaging.

8. Since aptamers are rapidly eliminated by the urinary pathway, we usually follow the biodistribution of aptamers until 3 h post-injection. However, longer kinetic could be performed in order to study the retention of aptamers in tumors.

9. Some fDOT systems include calibration curves in order to quantify the pmoles of dye in an ROI. Calibration curves could be obtained using the following procedure:

   (a) Make a set of five capillaries filled with different concentrations of fluorophores such that the amount is, respectively, 5, 10, 20, 40, and 80 pmol.

   (b) Insert one by one each capillary under the skin of an anesthetized nude mouse and proceed to 3D image acquisition. It is also possible to use a phantom made of epoxy resin and black ink miming mouse properties.

   (c) Reconstruct each 3D image and apply the previously described segmentation procedure. Plot the actual concentrations of capillaries according to the measured fluorescence intensity and estimate by linear regression the parameters of the affine calibration curve.

Acknowledgements

We are grateful to Dr. Rui Sousa (University of Texas, San Antonio) for his generous gift of a T7Y639F RNA polymerase-expressing plasmid and to Carine Pestourie, Agnès Cibiel, Benoit Jego, Isabelle Janssens, Daniel Miotto Dupont, Anikitos Garofalakis, and Bertrand Tavitian for their work on aptamer imaging in our laboratory. Studies relating to in vivo imaging of aptamers in our laboratories were supported by grants from the "Agence Nationale pour la Recherche" [projects ANR-RNTS TomoFluo3D, ANR-PNANO nanorings and under the frame of EuroNanoMed (project META)]; the FMT-XCT European program [Grant agreement no. 201792]; and the European Molecular Imaging Laboratory (EMIL) network [EU contract LSH-2004-503569].

References

1. Pestourie C, Tavitian B, Duconge F (2005) Aptamers against extracellular targets for in vivo applications. Biochimie 87:921–930

2. Cibiel A, Pestourie C, Duconge F (2012) In vivo uses of aptamers selected against cell surface biomarkers for therapy and molecular imaging. Biochimie 94:1595–1606

3. Keefe AD, Pai S, Ellington A (2010) Aptamers as therapeutics. Nat Rev Drug Discov 9:537–550

4. Hong H, Goel S, Zhang Y, Cai W (2011) Molecular imaging with nucleic acid aptamers. Curr Med Chem 18:4195–4205

5. Blank M, Weinschenk T, Priemer M, Schluesener H (2001) Systematic evolution of a DNA aptamer binding to rat brain tumor microvessels. selective targeting of endothelial regulatory protein pigpen. J Biol Chem 276:16464–16468

6. Stanlis KK, McIntosh JR (2003) Single-strand DNA aptamers as probes for protein localization in cells. J Histochem Cytochem 51:797–808

7. Li W, Yang X, Wang K, Tan W, He Y, Guo Q, Tang H, Liu J (2008) Real-time imaging of protein internalization using aptamer conjugates. Anal Chem 80(13):5002–5008

8. Xiao Z, Shangguan D, Cao Z, Fang X, Tan W (2008) Cell-specific internalization study of an aptamer from whole cell selection. Chemistry 14:1769–1775

9. Shi H, Tang Z, Kim Y, Nie H, Huang YF, He X, Deng K, Wang K, Tan W (2010) In vivo fluorescence imaging of tumors using molecular aptamers generated by cell-SELEX. Chem Asian J 5:2209–2213

10. Zueva E, Rubio LI, Duconge F, Tavitian B (2010) Metastasis-focused cell-based SELEX generates aptamers inhibiting cell migration and invasion. Int J Cancer 128:797–804

11. Cibiel A, Dupont DM, Duconge F (2011) Methods to identify aptamers against cell surface biomarkers. Pharmaceuticals 4:1216–1235

12. Shi H, He X, Wang K, Wu X, Ye X, Guo Q, Tan W, Qing Z, Yang X, Zhou B (2011) Activatable aptamer probe for contrast-enhanced in vivo cancer imaging based on cell membrane protein-triggered conformation alteration. Proc Natl Acad Sci 108(10):3900–3905

13. Shi H, Cui W, He X, Guo Q, Wang K, Ye X, Tang J (2013) Whole cell-SELEX aptamers for highly specific fluorescence molecular imaging of carcinomas in vivo. PLoS One 8, e70476

14. Cibiel A, Nguyen Quang N, Gombert K, Thézé B, Garofalakis A, Duconge F (2014) From ugly duckling to swan: unexpected identification from cell-SELEX of an anti-annexin a2 aptamer targeting tumors. PLoS One 9, e87002

15. Garofalakis A, Dubois A, Kuhnast B, Dupont DM, Janssens I, Mackiewicz N, Dolle F, Tavitian B, Duconge F (2010) In vivo validation of free-space fluorescence tomography using nuclear imaging. Opt Lett 35:3024–3026

16. Stuker F, Ripoll J, Rudin M (2011) Fluorescence molecular tomography: principles

and potential for pharmaceutical research. Pharmaceutics 3:229–274

17. Herve L, Koenig A, Da Silva A, Berger M, Boutet J, Dinten JM, Peltie P, Rizo P (2007) Noncontact fluorescence diffuse optical tomography of heterogeneous media. Appl Opt 46:4896–4906

18. Koenig A, Boutet J, Herve L, Berger M, Dinten JM, Da Silva A, Peltie P, Rizo P (2007) Fluorescence diffuse optical tomographic (fDOT) system for small animal studies. Conf Proc IEEE Eng Med Biol Soc 2007:2626–2629

19. Heetebrij RJ, Talman EG, v Velzen MA, van Gijlswijk RP, Snoeijers SS, Schalk M, Wiegant J, v d Rijke F, Kerkhoven RM, Raap AK et al (2003) Platinum(II)-based coordination compounds as nucleic acid labeling reagents: synthesis, reactivity, and applications in hybridization assays. Chembiochem 4:573–583

20. Tataurov AV, You Y, Owczarzy R (2008) Predicting ultraviolet spectrum of single stranded and double stranded deoxyribonucleic acids. Biophys Chem 133:66–70

21. Shigdar S, Qian C, Lv L, Pu C, Li Y, Li L, Marappan M, Lin J, Wang L, Duan W (2013) The use of sensitive chemical antibodies for diagnosis: detection of low levels of EpCAM in breast cancer. PLoS One 8, e57613

22. Rasband WS (1997–2014) ImageJ, U. S. National Institutes of Health, Bethesda, MD. http://imagej.nih.gov/ij/

23. Bezdek JC (1981) Pattern recognition with fuzzy objective function algorithms. Plenum, New York, NY

24. Dougherty R (2005) Extensions of DAMAS and benefits and limitations of deconvolution in beamforming. 11th AIAA/CEAS aeroacoustics conference. American Institute of Aeronautics and Astronautics, Monterey, CA. doi:10.2514/6.2005-2961.

25. Schneider CA, Rasband WS, Eliceiri KW (2012) NIH Image to ImageJ: 25 years of image analysis. Nat Methods 9:671–675

26. Bolte S, Cordelières FP (2006) A guided tour into subcellular colocalization analysis in light microscopy. J Microsc 224:213–232

27. Kamentsky L, Jones TR, Fraser A, Bray M-A, Logan DJ, Madden KL, Ljosa V, Rueden C, Eliceiri KW, Carpenter AE (2011) Improved structure, function and compatibility for cell profiler: modular high-throughput image analysis software. Bioinformatics (Oxford) 27:1179–1180

Chapter 12

Crystallographic Pursuit of a Protein-RNA Aptamer Complex

John J.G. Tesmer

Abstract

Only a few of the aptamers designed to selectively target proteins have been structurally characterized, such as those that target thrombin, von Willebrand factor, *Plasmodium falciparum* lactate dehydrogenase, interleukin 6, and platelet-derived growth factor B. Most of these aptamers are composed of DNA and were designed as therapeutics/diagnostics for targets found in human plasma. Recently, the crystal structure of a complex between an RNA aptamer and an intracellular target, G protein-coupled receptor kinase 2, was determined. Herein is described the overall approach used to isolate crystals that would allow the identification of the key interactions between aptamer and kinase. These strategies may be useful in structural characterization of other SELEX-generated RNA aptamer complexes.

Key words Crystallography, Aptamer, RNA, Complex, Inhibitor, Kinase, GRK2

1 Introduction

There are only a handful of aptamer complexes listed among the ~4400 crystal structures of protein–nucleic acid complexes currently deposited in the Protein Data Bank (PDB), despite the fact that aptamers have been developed for well over 1100 human proteins (www.somalogic.com). The paucity of protein–aptamer complexes in the PDB likely reflects inherent difficulties in crystallizing protein–nucleic acid complexes [1]. One of the chief hurdles is the conformational heterogeneity of nucleic acid, in particular that of RNA, which stems from its relatively weak tertiary interactions and ability to form a multitude of different base pairing and stacking interactions of comparable free energy. For these reasons, the nucleic acid components of these complexes often dominate crystal lattice contacts. Although such conformational plasticity may facilitate crystal formation, it often leads to imperfectly formed crystals. Another technical hurdle is the fact that aptamers tend to be non-globular, in contrast to proteins, and the resulting crystals exhibit high solvent content due to lattice packing constraints.

Günter Mayer (ed.), *Nucleic Acid Aptamers: Selection, Characterization, and Application*, Methods in Molecular Biology, vol. 1380, DOI 10.1007/978-1-4939-3197-2_12, © Springer Science+Business Media New York 2016

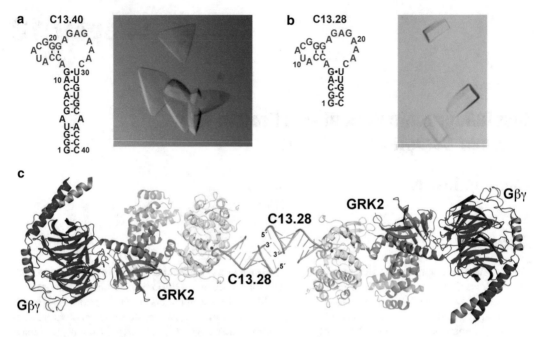

Fig. 1 GRK2-binding aptamers. (**a**) Secondary structure of the C13.40 aptamer, as revealed by structural studies. The nucleotides in *red* correspond to those that were altered by SELEX. The *inset* shows a representative crystal of the C13.40–GRK2–Gβγ complex, which diffracted to only 8.3 Å. (**b**) Secondary structure of the C13.28 aptamer. Note the altered stem nucleotides. The *inset* shows representative crystals of the C13.28–GRK2–Gβγ complex, which diffracted to 4.5 Å resolution. (**c**) Twofold crystal contact of the C13.28–GRK2–Gβγ complex mediated by base stacking at the 5′ end of the stem of the aptamer (*orange* phosphodiester backbone with bases represented as *cyan rods*). Mg^{2+} is shown as a *black sphere*

Both of these hurdles contribute to low diffraction resolution, and, in the worst case, to no crystals at all.

These hurdles were well evident in the case of the C13 RNA aptamer, developed via a SELEX approach to bind selectively to G protein-coupled receptor kinase 2 (GRK2) [2]. Crystals of GRK2 that diffracted to atomic resolution were only obtained after years of effort, and ultimately required the inclusion of heterotrimeric Gβγ subunits, which stabilize the kinase [3]. In comparison, the C13.40 variant of the C13 aptamer (Fig. 1a) crystallized in complex with GRK2-Gβγ nearly instantaneously, but the resulting crystals, albeit large, diffracted only to low resolution (8.3 Å). By the time the dust settled [4], over 12 unique crystal forms of GRK2 in complex with C13 aptamer variants were obtained. In each the RNA formed significant crystal lattice contacts wherein structural heterogeneity was apparent, and only the highest resolution structure allowed definition of the key interactions between the aptamer and the GRK2 active site.

Many previously reviewed techniques for optimization of protein–RNA crystals were used in pursuit of the GRK2–C13 structure, such as those covered in [1]. Thus, the focus of these procedures is on aptamer-specific crystallographic methods that proved helpful for the final successful structure determination. They may by extension be useful for the structural analysis of other protein–aptamer complexes.

2   Materials

2.1   Macromolecules

1. Aptamers: Ideally commercially synthesized and HPLC purified (e.g., by Integrated DNA Technologies) to help avoid RNase contamination, dissolved at 10 mM in DNase- and RNase-free ddH$_2$O (*see* **Note 1**).

2. Protein target(s): Purified to homogeneity, passed through a 0.2 μm spin filter, and concentrated typically to >1 mg/mL in size-exclusion chromatography (SEC) buffer (*see* below) (*see* **Note 2**).

2.2   SEC

1. SEC buffer: 20 mM HEPES pH 7–8 (but *see* **Note 1**), 50 mM NaCl, 5 mM MgCl$_2$, and 2 mM DTT (added fresh from a 1 M stock) (*see* **Note 3**).

2. Analytical SEC columns, such as S200 Superdex 10/300 (GE Healthcare) (*see* **Note 4**).

3. UV spectrophotometer to measure A$_{260}$ and A$_{280}$ readings from SEC fractions.

4. Protein concentrators (e.g., Amicon Ultra), ideally with molecular weight cutoffs below the molecular weight of the smallest species.

2.3   Crystallization Reagents

1. Commercial sparse matrix screens: The Natrix and Natrix 2 screens (Hampton Research) are a good starting point.

2. High-purity precipitants such as 30 % (w/v) stocks of PEG 3350 and 8000 (available from many vendors) and 1 M stocks of crystallization buffers in ddH$_2$O, all sterilized through 0.2 μM filters.

3. 96-Well sitting drop plates with three depressions for robotic screens (e.g., 96-3 well, INTELLI-PLATE® sold by Hampton Research). Sealing (packing) tape.

4. 24-Well hanging drop plates (e.g., VDX sold by Hampton Research) and siliconized cover slips.

3   Methods

3.1   Isolation and Evaluation of Protein–Aptamer Complexes

1. In a typical reaction, combine ~1 mg protein (e.g., GRK2, 80 kDa) in a 1:1.5 molar ratio with aptamer in 0.5 mL. Incubate on ice for 30 min (*see* **Note 5**). When Gβ$_1$γ$_2$ was included for co-crystallization, approximately 0.7 mg was also added and the mixture incubated on ice for an additional 30 min.

2. Inject no more than 0.5 mL of the sample onto the SEC column using an appropriate HPLC system. We use a BioLogic DuoFlow, but AKTA systems are also popular. Use a slow flow rate (~0.3 mL/min) and collect 300 µL fractions at 4 °C (*see* **Note 6**).

3. Determine peak protein fractions from A_{280} values. Using a standard curve (e.g., linear regression of log(Da) vs. (elution volume-void volume)), determine estimated molecular weight of the peaks (*see* **Note 7**).

4. Determine the A_{260}/A_{280} ratio in each peak estimated to have close to the correct complex molecular weight (*see* **Note 8**).

5. Pool fractions that have closest to the expected A_{260}/A_{280} ratio and/or most consistent ratios. Concentrate to 5–10 mg/mL (*see* **Note 9**).

6. Pass sample through a 0.2 µm spin filter prior to crystallization (*see* **Note 10**).

7. In the event that the aptamer affinity is too low or complex formation is ambiguous by SEC, it is reasonable to attempt to crystallize a complex of sample prepared by mixing the aptamer and protein together, and then proceeding directly to concentration and filtering. We had success via this approach using a 1:1.5 molar ratio of GRK2 to the C13.20 aptamer. However, there is always the danger that one might crystallize excess aptamer, necessitating proper controls (*see* Subheading 3.2, **step 1**).

3.2 Crystallization

There are many approaches to crystallizing macromolecules [5], but the traditional hanging drop method is one of the most robust. For example, the GRK2–aptamer complexes were typically mixed 1:1 (v/v) with well solution consisting of precipitant (such as polyethylene glycol (PEG) 3350 or 8000) in a 2 µL drop that was suspended over a 1 mL well. Incubation temperature is an important variable in screening for crystals, but for RNA aptamers 4 °C is the best initial choice because this may slow the rate of degradation and/or reduce conformational heterogeneity.

1. Set up a sparse matrix screen using various commercial kits and a crystallization robot (Art Robbins Gryphon is a popular choice). This allows one to assess a wide variety of conditions using smaller amounts of sample in shorter amounts of time than can be set up by hand. Robots such as this use a sitting drop format wherein there are typically three depressions for samples. Thus, for each well solution to be tested, one could screen three different protein concentrations; or two different protein concentrations and a buffer blank; or complex in one, aptamer in the second, and protein in the third. Thus one can easily control for (potentially) undesirable crystals that result from nonstoichiometric amounts of one component, from

components that dissociated under the crystallization conditions, or from buffer components.

2. Set up more focused homemade screens around promising conditions from the commercial screens or around those that are known to yield crystals of related samples. Below are standard initial grids we use in our studies. The recipes are for the well solutions (1 mL final volume) in a 24-well hanging drop plate (4 rows, 6 columns). Typically two trays of each grid are prepared so that one can be incubated at 4 °C and the other at 20 °C (*see* **Note 11**):

   *Grid 1: PEG 8000 versus pH, low salt (*see* Note 12):*

   30 % PEG 8000: 5 %, 10 %, 15 %, 20 % (w/v) (rows)

   pH: 100 mM buffer (pH 5, 5.5, 6, 6.5, 7, 7.5, 8, and 8.5) (columns) (*see* **Note 13**)

   Grid 2: PEG 8000 versus pH, high salt (NaCl):

   30 % PEG 8000: 5 %, 10 %, 15 %, 20 % (w/v) (rows)

   pH: 100 mM buffer (pH 5, 5.5, 6, 6.5, 7, 7.5, 8, and 8.5) (columns)

   NaCl: 1 M NaCl (e.g., 250 μL of 4 M in a 1.0 mL well) (all wells)

   Grid 3: PEG 3350 versus pH, low salt:

   30 % PEG 3350: 5 %, 10 %, 15 %, 20 % (rows)

   pH: 100 mM buffer (pH 5, 5.5, 6, 6.5, 7, 7.5, 8, and 8.5) (columns)

   Grid 4: PEG 3350 versus pH, high salt (NaCl):

   30 % PEG 3350: 5 %, 10 %, 15 %, 20 % (rows)

   pH: 100 mM buffer (pH 5, 5.5, 6, 6.5, 7, 7.5, 8, and 8.5) (columns)

   NaCl: 1 M NaCl (e.g., 250 μL of 4 M in a 1.0 mL well) (all wells)

   Grid 5: NaCl versus pH:

   4 M NaCl: 2 M, 2.5 M, 3 M, 3.5 M (rows)

   pH: 100 mM buffer (pH 5, 5.5, 6, 6.5, 7, 7.5, 8, and 8.5) (columns)

   Grid 6: Ammonium sulfate versus pH:

   3.5 M ammonium sulfate: 1.5 M, 2.0 M, 2.5 M, 3 M (rows) (*see* **Note 14**)

   pH: 100 mM buffer (pH 5, 5.5, 6, 6.5, 7, 7.5, 8, and 8.5) (columns)

   Grid 7: PEG 400 versus pH, low salt:

   100 % PEG 400: 20 %, 30 %, 40 %, 50 % (v/v) (rows)

pH: 100 mM buffer (pH 5, 5.5, 6, 6.5, 7, 7.5, 8, and 8.5) (columns)

Grid 8: PEG 400 versus pH, high salt:

100 % PEG 400: 20 %, 30 %, 40 %, 50 % (v/v) (rows)

pH: 100 mM buffer (pH 5, 5.5, 6, 6.5, 7, 7.5, 8, and 8.5) (columns)

NaCl: 1 M NaCl (e.g., 250 μL of 4 M in a 1.0 mL well) (all wells)

3.3 Improvement of Low-Quality Crystals

Although rapidly forming large trigonal plate crystals, the C13.40–GRK2–Gβ₁γ₂ complex had 86 % solvent content and diffracted to only 8.7 Å, which is insufficient for determining the molecular basis of aptamer selectivity, especially when the aptamer structure is unknown. There are fortunately many ways to improve crystal quality. Below are aptamer-specific steps we undertook in pursuit of our highest resolution structure at 3.5 Å spacings.

1. Trim unnecessary portions/projections of the aptamer. This will likely reduce flexibility and make the aptamer more globular, but may come at the price of lower affinity. For example, the C13.40 aptamer, which forms a relatively simple hairpin-loop/stem structure (Fig. 1a), made lattice contacts via end-to-end contacts by its flexible stem. Such contacts are frequently observed in RNA crystals [6]. The stem did not correspond to the selected region of the aptamer and thus was dispensable except for its potential role in stabilizing the fold of the aptamer. We therefore designed C13.28, wherein the stem was shortened by six base pairs compared to C13.40 (Fig. 1a, b). As hypothesized, C13.28 retained similar lattice packing to the C13.40 crystals and had reduced solvent content (83 %) and RNA flexibility because the end-to-end dimer contact was reduced by 12 base pairs, or approximately one full helical turn of A-form RNA. The diffraction limit was improved to 4.5 Å (Fig. 1c).

2. Engineer tighter junctions. For example, a 5′ G overhang was included in the C13.29 variant of the aptamer to try to stabilize the end-to end junction in the C13.28-GRK2-Gβ₁γ₂ crystals with a G-G base pair [6]. Sequences in the stem can also be optimized to contain more stable, GC-rich Watson-Crick base pairs, such as in the C13.28 variant (Fig. 1b), where 5′ GGCA 3′/3′ CCGU 5′ was introduced at the termini.

3. Eliminate promiscuous lattice forming motifs. Removal of non-essential portions of the aptamer that favor heterogeneous or flexible lattice contacts, such as the stem in C13.28 (Fig. 1c), is probably one of the most effective strategies. For example, the best diffracting crystals for the GRK2–aptamer

complex were obtained using C13.18, which contained only 18 bases of the 20 selected nucleotides of the original aptamer (56 % solvent content). A 3.5 Å structure was achieved, but at the price of 35-fold less binding affinity and tenfold lower IC_{50} against kinase activity.

4. Pray to a different deity, but meanwhile keep engineering and screening!

4   Notes

1. Although the C13 aptamer variants were very stable and did not require special consideration with regard to enzymatic degradation, it is advised that all solutions be treated so that they are RNase free (e.g., 0.1 % diethylpyrocarbonate, but therefore avoid using buffers with primary amines like Tris and HEPES, which inactivate this chemical). Bake all glassware (e.g., 300 °C for 2 h) that comes into contact with the working solutions. Designate an RNase-free workspace and clean this space regularly. Also designate a set of pipettes for use only in this area and use barrier tips. Wear a lab coat and gloves to prevent user contamination.

    A nucleic acid renaturation step may also be useful before attempting to isolate complexes with proteins. We did not perform this step for crystallographic work because we assumed mis-folded aptamers would not bind and be separated from properly folded RNA by SEC. However, renaturation could be required for more complex aptamers, or when one is trying to just "mix and go" (see Subheading 3.1, **step 7**). What follows is an example protocol:

    (a) Incubate 100 μM aptamer at 95 °C for 5 min in 20 mM Tris–HCl pH 7.0, 50 mM NaCl.

    (b) Incubate on ice for 2 min.

    (c) Supplemented to 5 mM $MgCl_2$ final concentration and incubate at 37 °C for 15 min.

2. Often proteins are buffer exchanged in a final polishing step using SEC. This can be omitted if the protein–aptamer complex is going to be purified via SEC because this will reduce loss of material.

3. The $MgCl_2$ is included as a counter ion for the nucleic acid phosphodiester backbone. When $G\beta_1\gamma_2$ was to be co-crystallized, the buffer also contained 10 mM CHAPS to stabilize the geranylgeranyl group at the C-terminus of the $G\gamma_2$ subunit.

4. Two columns connected in tandem will improve resolution of bound from unbound species, but may not be necessary if their molecular weight differences are large enough.

5. The strategy is to saturate the higher molecular species with the lower molecular weight one, driving all the higher molecular weight species into a complex. This allows for better separation from the unbound lower apparent molecular weight species, which could ultimately inhibit crystallization by "poisoning" the growing lattice. For small proteins or large aptamers, the protein may need to be the component in excess.

6. In principle, the smaller the injection volume, the better the peak resolution. Smaller amounts of material may also reduce peak broadening, but 250 µg of material is suggested at minimum because in all runs, up to 100 µg sample can be absorbed by the column, filters, and tubing.

7. Asymmetric protein complexes will often elute before the volume predicted for their molecular weight, and macromolecules that nonspecifically interact with the resin will elute after. Check for such aberrant behavior in a control run with protein (or aptamer) alone.

8. For example, the C13.40 RNA aptamer, a 40-nucleotide truncation of C13 (Fig. 1a), eluted as a complex with GRK2–$G\beta_1\gamma_2$ at ~140 kDa, consistent with its predicted molecular weight of ~140 kDa (126 kDa protein and 13 kDa RNA). The A_{260}/A_{280} ratio in peak fractions was 1.5, corresponding to ~85 % protein and ~15 % nucleic acid, close to the 90 % protein and 10 % RNA predicted if the molecules are in a stoichiometric complex [7].

9. There is no absolute rule of thumb for what concentration of complex is optimal for crystallization, because the ideal value is closely linked to the solubility of the complex being isolated. This parameter is often optimized during the crystallization process. For GRK2–aptamer complexes, we have had good luck in the range of 6–10 mg/mL.

10. It is advised to set up crystallization screens immediately, as opposed to freezing the complex in liquid N_2 for another day, because freeze thaw cycles likely damage samples. That said, it is probably a good idea to freeze away enough protein from each preparation so one can try to repeat a few crystallizations in the event that future preparations of the complex fail to produce crystals.

11. Keep the complex on ice whenever possible. When mixing complex with well solution on cover slips, add the complex to drop last because this will help draw the sample out of the pipette tip. If the drops are small, and if you are operating at room temperature (not advised for sensitive samples), one has to work quickly because the volume change that occurs via evaporation will be greater and irreproducible results may be

obtained. The cover slips bearing the sample are suspended over the well solution and sealed, typically with vacuum grease. When setting the cover slips, try not to let the cover slips overlap and be sure each well is sealed. Too much grease will lead to wandering cover slips and air gaps. Inspect each tray via microscope immediately after setup. Mark anything that could later be mistaken for a crystal in your notes. Check daily for 3 days, then every other day, then weekly, etc. If certain pH ranges or precipitant concentrations invariably precipitate the complex (heavy white snow-like material), adjust the grids accordingly to avoid these conditions and screen more useful chemical space. If the sample does not precipitate under any condition, try increasing either the protein or the precipitant concentration.

12. Many complexes are less soluble in solutions at low ionic strength. The low salt concentration may have to be adjusted upwards (e.g., to 100 mM) or the maximum precipitant concentration downwards in order to keep the complex in solution long enough for crystals to nucleate.

13. It is best to use buffers that have broad useful pH ranges because this reduces the number of parameters that vary across the tray. Good buffers are preferred [8], although Na-citrate is a very useful reagent for the lower pH ranges given its broad buffering range (3–6.2). Because Mg^{2+} will typically be included as a counter ion for the nucleic acid, phosphate buffers should be avoided because $MgPO_4$ forms beautiful but ultimately very disappointing crystals.

14. Ammonium sulfate stocks should be remade with more frequency than other precipitants. Stock solution could either be used as is (will be acidic) or individual stocks adjusted with NaOH to the designated pH of each column of the grid.

Acknowledgement

This work was supported by the US National Institutes of Health (NIH) grant HL086865.

References

1. Ke A, Doudna JA (2004) Crystallization of RNA and RNA-protein complexes. Methods 34(3):408–414

2. Mayer G, Wulffen B, Huber C et al (2008) An RNA molecule that specifically inhibits G-protein-coupled receptor kinase 2 in vitro. RNA 14(3):524–534

3. Lodowski DT, Pitcher JA, Capel WD et al (2003) Keeping G proteins at bay: a complex between G protein-coupled receptor kinase 2 and Gβγ. Science 300(5623):1256–1262

4. Tesmer VM, Lennarz S, Mayer G et al (2012) Molecular mechanism for inhibition of G protein-coupled receptor kinase 2 by a selective RNA aptamer. Structure 20(8):1300–1309. doi:10.1016/j.str.2012.05.002

5. McPherson A, Gavira JA (2014) Introduction to protein crystallization. Acta Crystallogr F Struct

Biol Commun 70(Pt 1):2–20. doi:10.1107/S2053230X13033141

6. Mooers BH (2009) Crystallographic studies of DNA and RNA. Methods 47(3):168–176. doi:10.1016/j.ymeth.2008.09.006, S1046-2023(08)00152-7 [pii]

7. Sambrook J, Russell DW (2001) Molecular cloning: a laboratory manual. Cold Spring Harbor Laboratory Press, Cold Spring Harbor

8. Ferguson WJ, Braunschweiger KI, Braunschweiger WR et al (1980) Hydrogen ion buffers for biological research. Anal Biochem 104(2):300–310

Part III

Application

Chapter 13

Voltammetric Aptasensor Based on Magnetic Beads Assay for Detection of Human Activated Protein C

Arzum Erdem, Gulsah Congur, and Ece Eksin

Abstract

Aptamers are defined as new generation of nucleic acids, which has recently presented promising specifications over to antibodies. An increasing number of electrochemical studies related to aptamer-based sensors, so-called aptasensors have been introduced in the literature. Herein, the interaction between human activated protein C (APC) and its cognate DNA aptamer (DNA APT) was performed at the surface of magnetic beads (MBs), followed by voltammetric detection using disposable graphite electrodes (PGEs).

Key words Aptasensors, Electrochemistry, Voltammetry, Human activated protein C, Magnetic beads

1 Introduction

Aptamers are nucleic acid ligands, which are isolated from a synthetic nucleic acid pool by *S*ystematic *E*volution of *L*igands by *EX*ponential enrichment (SELEX) [1–3]. They recognize their target molecules, e.g., amino acids, drugs, and proteins, with high affinity and specificity. Many reports evaluated aptamer (APT)-based detection assays in combination with optical [4–6], piezoelectrical [7–10], and electrochemical techniques [3, 11–22].

Activated protein C (APC) is a serine protease and the key enzyme of the protein C (PC) pathway [11, 17, 19, 23–26]. APC has crucial properties for protection of endothelial barrier function, whereas APC resistance is a lifelong effected process and results in deregulation of the PC pathway. Recombinant APC has been used as a therapeutic for sepsis treatment. Several studies report APC detection by ELISA [26] and fluorometry [27].

In this protocol, the interaction between APC and its cognate DNA aptamer (DNA APT) at the surface of magnetic beads (MB) is demonstrated and we employ voltammetric detection by using disposable graphite electrode (Scheme 1). First, APC was

Günter Mayer (ed.), *Nucleic Acid Aptamers: Selection, Characterization, and Application*, Methods in Molecular Biology, vol. 1380, DOI 10.1007/978-1-4939-3197-2_13, © Springer Science+Business Media New York 2016

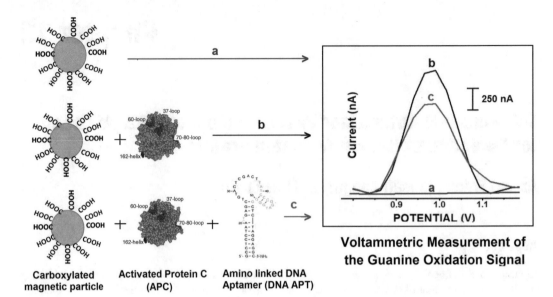

Carboxylated magnetic particle   Activated Protein C (APC)   Amino linked DNA Aptamer (DNA APT)

Scheme 1 Experimental scheme for electrochemical detection of interaction between APC and its DNA aptamer. DPVs representing the (a) control signal, the guanine signal obtained in the presence of (b) DNA APT and (c) after interaction of APC and DNA APT

immobilized on the surface of carboxylated MB and second, the interaction in the presence of amino-modified DNA APT was monitored based on the changes at the guanine oxidation signal measured by differential pulse voltammetry (DPV). This protocol describes the accurate, reliable, and fast analysis of APC.

2   Materials

2.1   Magnetic Separation Components

1. *Magnetic separator.* Use a magnetic separator MCB 1200 for each mixing/separation step unless otherwise specified.

2. *Magnetic particles (MBs):* The carboxylated magnetic beads (MB-COOH) of 0.44 mm diameter (Estapor, Merck). Store the MBs at 4 °C.

Electrochemical Assay Components (Three-Electrode System) and Measurements [3, 18–20, 28, 29]

1. *Pencil graphite electrodes (PGEs):* 0.5 HB TOMBOW pencil leads. The length of a pencil lead is 6 mm. Cut it in half as 3 mm. Use a Rotring pencil to hold 1.4 mm of the lead.

2. *Reference electrode:* Ag/ AgCl/ 3 M KCl (BAS, ModelRE-5B).

3. *Auxiliary electrode:* a platinum wire.

4. *Measurements.* The oxidation signal of guanine can be measured by using Autolab PGSTAT electrochemical analysis system and GPES 4.9007 software package (EcoChemie) using differential pulse voltammetry (DPV) technique following the treatment of the data by Savitzky and Golay filter (level 2) and the moving average baseline correction with a "peak width" of 0.03.

2.2 Preparation of the Solutions

1. *Washing solution of MBs*: 5 mM Tris–HCl, 20 mM NaCl, pH 7.0. Weigh 197 mg Tris–HCl and transfer in a glass baker containing 230 mL water. Add about 20 mL 1 M NaOH solution, mix, and adjust pH with HCl. Store at +4 °C (*see* **Note 1**).

2. *Preparation solution of APC and APC-specific DNA aptamer*: 50 mM phosphate buffer, 20 mM NaCl, pH 7.4. Weigh 1.36 g KH_2PO_4, 6.96 g K_2HPO_4 and 1.168 g NaCl and transfer them in a glass baker containing 900 mL water (*see* **Note 1**). Then, add 1 M NaOH, mix, and adjust pH. Fill-up to 1 L with H_2O. Store at +4 °C.

3. *Washing solution of MBs after APC immobilization and the interaction of APC and its DNA aptamer (DNA APT)*: 10 mM phosphate buffer, 150 mM NaCl, 3 mM $MgCl_2$, pH 7.4 (PBSMg). Weigh 0.272 g KH_2PO_4, 1.392 g K_2HPO_4, 2.19 g NaCl and 0.0715 g $MgCl_2$ and transfer them in a glass baker containing 230 mL water (*see* **Note 1**). Add 1 M NaOH, mix and adjust pH. Make up to 250 mL with H_2O. Store at +4 °C.

4. *Preparation solution of APC-specific DNA aptamer*: 5 mM Tris–HCl, 20 mM NaCl, pH 7.0 (TBS). Store at +4 °C.

5. *Alkaline solution*: 0.02 M NaOH. Weigh 20 g NaOH and dissolve 0.5 L water to prepare 1 M NaOH (*see* **Note 1**). Then, dilute at 0.02 M in H_2O.

6. *Diluted solution of the samples, pretreatment, and measurement solution of PGEs*: 0.5 M acetate buffer, 20 mM NaCl, pH 4.8 (ABS). Add 28.9 mL concentrated acetic acid solution into a glass baker containing 500 mL water (*see* **Note 1**). Transfer 1.168 g NaCl and mix. Add 1 M NaOH solution to adjust pH. Fill up to 1 L with H_2O. Store at +4 °C.

3 Methods

Perform all procedures at room temperature unless otherwise indicated.

3.1 Immobilization of APC at the Surface of MB

1. Transfer 3 μL MB into a 1.5 mL centrifuge tube. Add 90 μL TBS, pH 7.0 and wash the MBs by gentle mixing (*see* **Note 2**).

2. Separate the MB from the solution by using magnetic separator and remove the solution.

3. Add 30 μL of required concentration of APC solution and gently mix for 15 min (*see* **Notes 3** and **4**).

4. Separate the MBs as **step 2**.

5. Wash the APC immobilized MBs with 90 μL PBSMg.

6. Separate the MBs as **step 2**.

3.2 Interaction of APC with Its DNA APT at the Surface of MB

1. Add 30 µL of 100 µg/mL DNA APT at the surface of APC-immobilized MBs and incubate during 15 min with gentle mixing (*see* **Notes 5** and **6**).

2. Wash the MBs with 90 µL PBSMg.

3. Separate the MBs from the solution by using magnetic separator and remove the solution.

3.3 Alkaline Treatment

1. Add 25 µL of 0.02 M NaOH into the vials containing MBs and incubate during 5 min with gentle mixing.

2. Use the magnetic separator to remove the samples.

3. Transfer the samples into the vials containing 85 µL ABS, pH 4.8. Total sample volume is 110 µL.

3.4 Immobilization of the Samples at the Surface of Disposable PGEs

1. Vortex the vials containing 110 µL sample for 1 min.

2. Immerse the pretreated PGEs into the vials and wait for 15 min in order to complete passive adsorption process.

3. Wash the PGEs into the ABS (pH 4.8) during 5 s.

4. Transfer the PGEs into three electrochemical cells.

3.5 Electrochemical Measurements

1. Adjust the scanning potential range from +0.2 V to +1.45 V.

2. Use the pulse amplitude as 50 mV with the scan rate as 50 mV/s.

3. Measure the guanine oxidation signal by using DPV in ABS (pH 4.8) (*see* **Note 7**) [30].

---

4   Notes

1. Prepare all solutions in ultrapure water (deionized water with the sensitivity of 18 MΩ cm at 25 °C). Other chemicals are of analytical reagent grade.

2. The control measurements were performed in the presence of MB (Fig. 1A-a).

3. 5, 15, and 30 µg/mL APC was immobilized at the surface of MBs and the specific interaction of APC and 3′ amino-labeled DNA APT (Fig. 2).

4. The control measurements were performed in the presence of MB and APC at its different concentrations (Fig. 2A-a).

5. As control, the DNA APT was immobilized at the surface of MB without APC. The guanine signal was measured after the immobilization of 100 µg/mL DNA APT which was labelled with amino group at 3′ (Fig. 1A-b) or 5′ end (Fig. 1A-c).

6. The average guanine signals were measured as 1261.0 ± 243.3 nA and 978 ± 62.2 nA with the relative standard deviation (RSD) % as 19.3 % and 6.4 % ($n = 3$) in the presence of 3′ (Fig. 1B-a) or

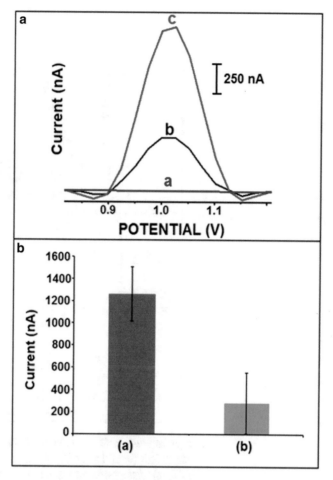

Fig. 1 (**a**) *Voltammograms* representing the oxidation signals of guanine obtained before (control signal, a), and after 100 μg/mL 5′ (b) or 3′ (c) amino linked DNA APT immobilization onto the surfaces of MBs. (**b**) *Histograms* representing the average guanine signal (*n* = 3) in the presence of 100 μg/mL 5′ (a) or 3′ (b) amino-linked DNA APT

5′ DNA APT (Fig. 1B-b). The 3′ amino-labeled DNA APT was chosen for further studies due to the highest guanine signal was measured.

7. After interaction of 5 and 30 μg/mL APC (Fig. 2B-a and c, respectively), 13.6 % and 33.2 % increase at guanine signal was observed whereas 19.3 % decrease was obtained in the presence of 15 μg/mL APC. After the interaction of 15 μg/mL APC and 100 μg/mL DNA APT the average guanine signal was measured as 1095.3±213 nA (RSD % = 19.4 %, *n* = 3). It was concluded that the specific interaction was occurred at 15 μg/mL APC concentration due to the decrease at the guanine signal could be obtained similar to our previous study [19].

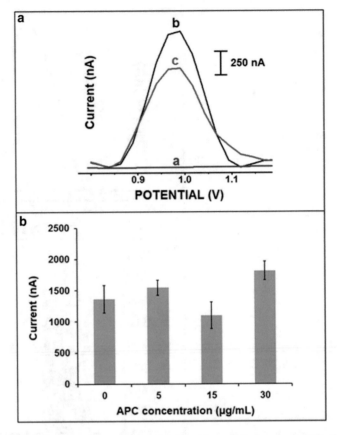

Fig. 2 (a) *Voltammograms* representing the (a) control signal, guanine signals obtained before (b) and after (c) interaction of 15 μg/mL APC and 100 μg/mL DNA APT. (b) *Histograms* representing average the guanine signals obtained before and after interaction of APC at its different concentrations and 100 μg/mL DNA APT

Acknowledgement

A.E. acknowledges the financial support from Turkish Scientific and Technological Research Council (TUBITAK) through the grant (Project No.111T073). G.C. acknowledges, respectively, the project scholarship by TUBITAK (Project No.111T073).

References

1. Tuerk C, Gold L (1990) Systematic evolution of ligands by exponential enrichment: RNA ligands to bacteriophage T4 DNA polymerase. Science 249:505–510

2. Famulok M, Hartig JS, Mayer G (2007) Functional aptamers and aptazymes in biotechnology, diagnostics, and therapy. Chem Rev 107:3715–3743

3. Erdem A, Karadeniz H, Mayer G, Famulok M, Caliskan A (2009) Electrochemical sensing of aptamer-protein interactions using a magnetic particle assay and single-

use sensor technology. Electroanalysis 21: 1278–1284

4. Wang R, Xiang Y, Zhou X, Liu L, Shi H (2015) A reusable aptamer-based evanescent wave all-fiber biosensor for highly sensitive detection of Ochratoxin A. Biosens Bioelectron 66:11–18

5. Li W, Dong Y, Wang X, Li H, Xu D (2015) PolyA-tailed and fluorophore-labeled aptamer-gold nanoparticle conjugate for fluorescence turn-on bioassay using iodide-induced ligand displacement. Biosens Bioelectron 66:43–49

6. Feng C, Dai S, Wang L (2014) Optical aptasensors for quantitative detection of small biomolecules: a review. Biosens Bioelectron 59:64–74

7. Tombelli S, Minunni M, Luzi E, Mascini M (2005) Aptamer-based biosensors for the detection of HIV-1 Tat protein. Bioelectrochemistry 67:135–141

8. Tombelli S, Bini A, Minunni M, Macsini M (2009) Piezoelectric biosensors for aptamer—protein interaction, biosensors and biodetection. Methods Mol Biol 504:23–36

9. Minunni M, Tombelli S, Gullotto A, Luzi E, Mascini M (2004) Development of biosensors with aptamers as bio-recognition element: the case of HIV-1 Tat protein. Biosens Bioelectron 20:1149–1156

10. Liss M, Petersen B, Wolf H, Prohaska E (2002) An aptamer-based quartz crystal protein biosensor. Anal Chem 74:4488–4495

11. Erdem A, Congur G, Mese F (2014) Electrochemical detection of activated protein C using an aptasensor based on PAMAM dendrimer modified pencil graphite electrodes. Electroanalysis 26:1–12

12. Wang W, Chen C, Qian M, Zhao XS (2008) Aptamer biosensor for protein detection using gold nanoparticles. Anal Biochem 373: 213–219

13. Park MK, Kee JS, Quah JY, Netto V, Song J, Fang Q, Fosse ME, Lo GQ (2013) Label-free aptamer sensor based on silicon microring resonators. Sensor Actuator B 176:552–559

14. Palchetti I, Mascini M (2012) Electrochemical nanomaterial-based nucleic acid aptasensors. Anal Bioanal Chem 402:3103–3114

15. Congur, G.; Erdem, A. Micro and nanopatterning for bacteria and virus-based biosensing applications. In *Portable Biosensing of Food Toxicants and Environmental Pollutants*; Nikolelis, D.P., Varzakas, T., Erdem, A., Nikoleli, G.-P., Ed.; CRC Press: Boca Raton, FL, USA, 2013, pp. 681–694

16. Rodriguez MC, Kawde AN, Wang J (2005) Aptamer biosensor for label-free impedance

17. Erdem A, Congur G (2014) Dendrimer enriched single-use aptasensor for impedimetric detection of activated protein C. Colloids Surf B Biointerfaces 117:338–345

18. Erdem A, Congur G (2014) Dendrimer modified 8-channel screen-printed electrochemical array system for impedimetric detection of activated protein C. Sensor Actuator B 196:168–174

19. Erdem A, Congur G (2014) Voltammetric aptasensor combined with magnetic beads assay developed for detection of human activated protein C. Talanta 128:428–433

20. Erdem A, Eksin E, Muti M (2014) Chitosan–graphene oxide based aptasensor for the impedimetric detection of lysozyme. Colloid Surf B 115:205–211

21. Rohrbach F, Karadeniz H, Erdem A, Famulok M, Mayer G (2012) Label-free impedimetric aptasensor for lysozyme detection based on carbon nanotube-modified screen-printed electrodes. Anal Biochem 421:454–459

22. Kawde AN, Rodriguez MC, Lee TMH, Wang J (2005) Label-free bioelectronic detection of aptamer–protein interactions. Electrochem Commun 7:537–540

23. Esmon CT (2003) The protein C pathway. Chest 124:26S–32S

24. Dahlback B, Villoutreix BO (2005) Regulation of blood coagulation by the protein C anticoagulant pathway novel insights into structure–function relationships and molecular recognition. Arterioscler Thromb Vasc Biol 25:1311–1320

25. Muller J, Isermann B, Ducker C, Salehi M, Meyer M, Friedrich M, Madhusudhan T, Oldenburg J, Mayer G, Potzsch B (2009) An exosite-specific ssDNA aptamer inhibits the anticoagulant functions of activated protein C and enhances inhibition by protein C inhibitor. Chem Biol 16:442–451

26. Muller J, Friedrich M, Becher T, Braunstein J, Kupper T, Berdel P, Gravius S, Rohrbach F, Oldenburg J, Mayer G, Potzsch BJ (2012) Monitoring of plasma levels of activated protein C using a clinically applicable oligonucleotide-based enzyme capture 1 assay. Journal of Thrombosis and Haemostasis 10:1–9

27. Zhao Q, Gao J (2013) Fluorogenic assays for activated protein C using aptamer modified magnetic beads. Microchim Acta 180:813–819

28. Wang J, Kawde AN, Erdem A, Salazar M (2001) Magnetic bead-based label-free electrochemical detection of DNA hybridization. Analyst 126:2020–2024

29. Erdem A, Pividori MI, Lermo A, Bonanni A, delValle M, Alegret S (2006) Genomagnetic assay based on label-free electrochemical detection using magneto-composite electrodes. Sensor Actuator B 114:591–598

30. Gulmez B, Karadeniz H, Erdem A, Ozsoz M (2007) Electrochemical sensing of calf thymus double stranded DNA and single stranded DNA by using a disposable sensor. Compr Anal Chem 49:e195–e202

Chapter 14

Apta-PCR

Alessandro Pinto, Pedro Nadal Polo, Miriam Jauest Rubio,
Marketa Svobodova, Teresa Mairal Lerga, and Ciara K. O'Sullivan

Abstract

Real-time Apta-PCR is a methodology that can be used for a wide variety of applications ranging from food quality control to clinical diagnostics. This method takes advantage of the combination of the sensitivity of nucleic acid amplification with the selectivity of aptamers. Ultra-low detection of target analyte can potentially be achieved, or, improved detection limits can be achieved with aptamers of low-medium affinity. Herein, we describe a generic methodology coined real-time Apta-PCR, using a model target (β-conglutin) and a competitive format, which can be adapted for the detection of any target which an aptamer has been selected for.

Key words Apta-PCR, Aptamer, Protein quantification

1 Introduction

The quantitative polymerase chain reaction (qPCR) facilitates the specific and sensitive real-time detection of nucleic acid molecules, whilst enzyme-linked immunosorbent assays (ELISA) or radio-immunoassays (RIA) are used for the quantitative detection of a wide range of protein targets. In the early 1990s, Cantor's group reported the exploitation of the quantitative amplification achievable with qPCR for the detection of a protein target in a technique he called Immuno-PCR [1], where specific antibody receptors were conjugated to DNA tags, that following formation of a sandwich-type immunocomplex served as a template for the qPCR reaction. Immuno-PCR offers great sensitivity, enhancing the sensitivity of a common immunoassay by up to 100,000-fold [2–5]; the technique suffers from some drawbacks related to the coupling of the antibody with the DNA reporter moieties. Such coupling often results in an uneven numbers of reporters per antibody, thus hindering true quantification [6, 7]. Additionally, following the immuno-recognition step, the DNA needs to be eluted from the antibody for subsequent amplification.

Günter Mayer (ed.), *Nucleic Acid Aptamers: Selection, Characterization, and Application*, Methods in Molecular Biology, vol. 1380, DOI 10.1007/978-1-4939-3197-2_14, © Springer Science+Business Media New York 2016

The real-time Apta-PCR described herein represents a further advancement and simplification of Immuno-PCR, where a single aptamer molecule replaces the cumbersome antibody-DNA complex. Recently, the aptamer specificity has been combined with qPCR sensitivity in various approaches for the ultrasensitive detection of proteins, including a proximity ligation assay [8–12], nuclease protection assay [13], capillary electrophoresis (CE) [14], and the use of target-modified magnetic microparticles [15]. In contrast to these reported approaches, real-time Apta-PCR does not rely on previous knowledge of the aptamer receptor and the instruments and methods used are widely available. The only requirement of the method is that the reporter aptamer to be employed is flanked with primer regions, and as the vast majority of reported aptamers have been selected by SELEX, they originally were flanked by primer-binding sites and this does not represent an issue—even in the scenario where part or all of the primer is involved in target binding, the aptamer will be released from the aptamer-target complex prior to amplification [16, 17]. Similar to ELISA, different types of real-time Apta-PCR can be performed, including direct or indirect sandwich and competitive assays.

In sandwich real-time Apta-PCR a capture affinity molecule, such as an aptamer or antibody is immobilized to capture the analyte of interest, which is effectively sandwiched between a capture biomolecule and a reporter aptamer. This format can provide better specificity as each of the capture and reporter biomolecules bind to a specific region on the analyte of interest. Competitive real-time Apta-PCR exploits competition of the analyte of interest with an immobilized standard amount of analyte for binding to the aptamer [18–20].

2 Materials

2.1 Plate Preparation

1. Immobilization buffer: 50 mM Carbonate/bicarbonate buffer: 44 mM sodium bicarbonate, 6 mM sodium carbonate, pH 9.6 (Sigma-Aldrich). Store at 4 °C.

2. Blocking buffer (PBS-Tween): 10 mM phosphate, 138 mM NaCl, 2.7 mM KCl, pH 7.4, 0.05 % v/v Tween-2® (see Note 1). Store at 4 °C.

3. Model target protein: β-Conglutin (Extracted, isolated and characterized as previously described [21]). Store at –20 °C.

4. Flat-bottomed microtiter plate.

2.2 Competitive Apta-PCR Assay

1. SELEX binding buffer (see Note 2): Phosphate-buffered saline (PBS) + 1.5 mM MgCl₂. Adjust to pH 7.4 with HCl and NaOH. Store at room temperature.

2. Model aptamer sequence: β-Conglutin binding aptamer (β-CBA) 5′-agc tga cac agc agg ttg gtg ggg gtg gct tcc agt tgg gtt gac aat acg tag gga cac gaa gtc caa cca cga gtc gag caa tct cga aat-3′. Store at –20 °C.

3. Salmon sperm DNA. Store at –20 °C.

4. MilliQ water (18.2 MΩ·cm).

2.3 Real-Time PCR

1. 10× PCR buffer: 200 mM Tris–HCl, 500 mM KCl. Adjust to pH 8.4 with HCl and NaOH. Store at –20 °C.

2. 50 mM $MgCl_2$. Store at –20 °C.

3. 2 mM dNTPs. Store at –20°C.

4. 10,000-fold concentrate SYBR Green I in DMSO (*see* **Note 3**). Store at –20 °C.

5. 1 μM ROX (*see* **Note 4**). Store at –20 °C.

6. 10 μM forward primer: 5′agc tga cac agc agg ttg gtg 3′. Store at –20 °C.

7. 10 μM reverse primer: 5′att tcg aga ttc ctc gac tcg tg 3′. Store at –20 °C.

8. Taq DNA polymerase. Store at –20 °C.

9. 96/384-well plates for qPCR.

10. Sealing film.

3 Methods

3.1 Plate Preparation

1. Dissolve the protein in immobilization buffer to a final concentration of 20 μg/mL.

2. Incubate 50 μL/well of the protein solution on each well of the microtiter plate for 30 min. at 37 °C under shaking conditions.

3. Remove the supernatant.

4. Wash three times with 200 μL of PBS-Tween buffer.

5. Incubate 200 μL/well of blocking buffer within the microtiter plate for 30 min. at 37 °C under shaking conditions.

6. Discharge the supernatant.

7. Wash three times with 200 μL of PBS-Tween buffer (*see* **Note 5**).

3.2 Competitive Apta-PCR Assay

1. In individual Eppendorf tubes, prepare a series of dilutions (0–1 μM) of the protein of interest (*see* **Note 6**) with 1 nM of aptamer (*see* **Note 7**), in the presence of 3 μg/mL salmon sperm DNA from baker's yeast in SELEX binding buffer. All samples including standards are carried out in triplicate.

2. Incubate this mixture solution for 30 min at 37 °C under moderate shacking condition (*see* **Note 8**).

3. Carefully transfer the 50 μL of this solution into individual wells of the previously coated microtiter plate.

4. Incubate the plate for 30 min at 37 °C under shaking conditions.

5. Carefully remove the supernatant.

6. Wash the plate three times with 200 μL of SELEX binding buffer and remove the supernatant.

7. Add 50 μL of MilliQ water.

8. Incubate the plate at 95 °C for 5 min.

9. Recover the supernatant containing the eluted aptamer.

3.3   Real-Time PCR

1. Prepare 1× PCR Master Mix on ice for desired amount of samples (*see* **Note 9**):

   Master Mix solutions for ten samples

   (a) 10 μL 10× PCR buffer

   (b) 10 μL $MgCl_2$

   (c) 10 μL 0.2 dNTPs

   (d) 1 μL of 1 μM ROX dye

   (e) 1 μL of forward primer

   (f) 1 μL of reverse primer

   (g) 1 μL of water solution of SYBR Green I 1–100

   (h) Taq enzyme 1 μL

   (i) Water to 100 μL

2. Transfer 1 μL of the eluted aptamer into individual wells of a 384 qPCR plate (*see* **Note 10**). Triplicate repeats are recommended. Three non-template controls (NTC) consisting of 1 μL of MilliQ water should also be included.

3. Using a multichannel pipette, transfer 9 μL 1× Master Mix (on ice) to each well.

4. Carefully seal the plate with the sealing film.

5. Place the plate in a qPCR instrument, previously programmed (*see* **Note 11**):

   (a) Denaturation:
       95 °C for 2 min.

   (b) 40× PCR cycles:
       95 °C for 10 s
       58 °C for 30 s
       72 °C for 30 s.

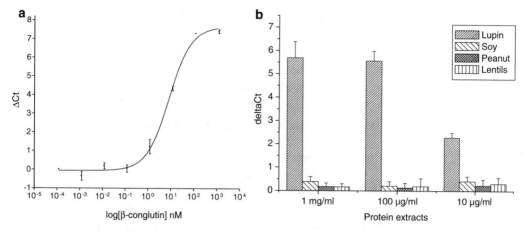

Fig. 1 Real-time Apta-PCR detection of β-conglutin. (**a**) Raw dates obtained from real-time PCR experiment. The numbers (1–8) indicate Ct values of different β-conglutin concentration from 0-365 μg/mL, 1:10 dilutions. (**b**) Calibration curve obtained for β-conglutin via real-time Apta-PCR competitive assay. Calibration curve was obtained by plotting the ΔCt values versus the concentration of β-conglutin, obtaining a linear range from 1.3 (0.382 μg/mL) to 68.4 nM (20.1 μg/mL) of β-conglutin with an EC_{50} value of 8.38 (2.5 μg/mL) and a L.O.D. of 85 pM (25 ng/mL). Inset: Cross-reactivity studies with soy, peanut and lentils. The error bars represent the standard deviation of 5 repetitions [18]

(c) Melting curve analysis:

95 °C for 15 s.

60 °C for 15 s.

95 °C for 15 s.

6. Analyze results using real-time PCR software to calculate the amount of aptamer in the sample, which, in turn, is proportional to the amount of analyte present.

7. Calculate ΔCt, which is the difference between the cycle threshold of the sample and the blank (sample containing aptamer not exposed to the target) (Fig. 1).

4   Notes

1. Buffers should be chosen according to their compatibility with the protein of interest. All solutions are prepared with deionized water purified with a resistivity of 18.2 MΩ·cm at 25 °C.

2. To achieve better performances, the buffer normally used is the one in which the aptamer was selected.

3. Pre-formulated Master Mix containing all the components (except template, primers and water) are commercially available.

4. ROX is a passive reference dye specific for the qPCR instrument used. Some other brands might use different dyes and the last generation of qPCR instruments do not need this internal reference.

5. Normally, plates filled with 200 μL of PBS-Tween can be stored at 4 °C up to 1 week, but this might vary according to the protein of interest.

6. From this step on, filtered tips should be used as they minimize the risk of contamination.

7. We have noticed that this particular concentration works consistently for different targets, giving controls, where the protein is absent, with an amplification profile equal to the one from the nontemplate control (the NTC and the no-protein control should have the same Ct, threshold Cycle). If the negative control differs from the NTC having a lower Ct, the concentration of the aptamer needs to be lowered.

8. If pre-prepared plates are used, during this step the microtiter plate should be equilibrated at 37 °C during this step. In case the microtiter plate are freshly prepared, this step is carried out at the same time as the **step 5** of the microtiter plate preparation.

9. Without the enzyme the buffer can be safely stored for a couple of weeks in freezer once prepared. The primer and the magnesium concentrations of this master mix have been optimised for β-conglutin detection but may vary for other targets. Other formulations containing DMSO, BSA, betaine, or other reagents can be used to improve the performance of the reaction.

10. In case of low reproducibility, higher amount of eluted DNA can be used for amplification: 10 μL of eluted DNA is transferred to individual wells of 96 qPCR plate with 10 μL of 2× PCR Master Mix to obtain a final volume of 20 μL of PCR product.

11. The PCR protocol should be selected according to the enzyme used.

References

1. Sano S, Smith CL, Cantor CR (1992) Immuno-PCR: very sensitive antigen detection by means of specific antibody-DNA conjugates. Science 258:120–122

2. Nam JM, Stoeva SI, Mirkin CA (2004) Bio-bar-code-based DNA detection with PCR-like sensitivity. J Am Chem Soc 126:5932–5933

3. Mweene AS, Ito T, Okazaki K, Ono E, Shimizu Y, Kida H (1996) Development of immuno-PCR for diagnosis of bovine herpesvirus 1 infection. J Clin Microbiol 34:748–750

4. Niemeyer CM, Adler M, Wacker R (2005) Immuno-PCR: high sensitivity detection of proteins by nucleic acid amplification. Trends Biotechnol 23:208–216

5. Adler M (2005) Immuno-PCR as a clinical laboratory tool. Adv Clin Chem 39:239–292

6. McKie A, Samuel D, Cohen B, Saunders NA (2002) Development of a quantitative immuno-PCR assay and its use to detect mumps-specific IgG in serum. J Immunol Methods 261:167–175

7. Niemeyer CM, Adler M, Pignataro B, Lenhert S, Gao S, Chi L et al (1999) Self-assembly of DNA-streptavidin nanostructures and their use as reagents in immuno-PCR. Nucleic Acids Res 27:4553–4561

8. Fredriksson S, Gullberg M, Jarvius J, Olsson C, Pietras K, Gústafsdóttir SM et al (2002) Protein detection using proximity-dependent DNA ligation assays. Nat Biotechnol 20:473–477

9. Gustafsdottir SM, Schallmeiner E, Fredriksson S, Gullberg M, Söderberg O, Jarvius M et al (2005) Proximity ligation assays for sensitive and specific protein analyses. Anal Biochem 345:2–9

10. Gustafsdottir SM, Nordengrahn A, Fredriksson S, Wallgren P, Rivera E, Schallmeiner E et al (2006) Detection of individual microbial pathogens by proximity ligation. Clin Chem 52:1152–1160

11. Yang L, Fung CW, Eun JC, Ellington AD (2007) Real-time rolling circle amplification for protein detection. Anal Chem 79:3320–3329

12. Yang L, Ellington AD (2008) Real-time PCR detection of protein analytes with conformation-switching aptamers. Anal Biochem 380:164–173

13. Wang XL, Li F, Su YH, Sun X, Li XB, Schluesener HJ et al (2004) Ultrasensitive detection of protein using an aptamer-based exonuclease protection assay. Anal Chem 76:5605–5610

14. Zhang H, Wang Z, Li XF, Le XC (2006) Ultrasensitive detection of proteins by amplification of affinity aptamers. Angew Chem Int Ed Engl 45:1576–1580

15. Fischer NO, Tarasow TM, Tok JBH (2008) Protein detection via direct enzymatic amplification of short DNA aptamers. Anal Biochem 373:121–128

16. Pinto A, Bermudo Redondo MC, Ozalp VC, O'Sullivan CK (2009) Real-time apta-PCR for 20 000-fold improvement in detection limit. Mol BioSyst 5(5):548–553

17. Pinto A, Lennarz S, Rodrigues-Correia A, Heckel A, O'Sullivan CK, Mayer G (2012) Functional detection of proteins by caged aptamers. ACS Chem Biol 7:359–365

18. Svobodova M, Miairal T, Nadal P, Bermudo Redondo MC, O'Sullivan CK (2014) Ultrasensitive aptamer based detection of β-conglutin food allergen. Food Chem 165:419–423

19. Pinto A, Nadal P, Henry O, Bermudo Redondo MC, Svobodova M, O'Sullivan CK (2014) Label-free detection of gliadin food allergen mediated by real-time apta PCR. Anal Bioanal Chem 406:515–524

20. Svobodova M, Bunka D, Nadal P, Stockley PG, O'Sullivan CK (2013) Selection of 2'F-modified RNA aptamers against prostate-specific antigen and their evaluation for diagnostic and therapeutic applications. Anal Bioanal Chem 405:9149–9157

21. Nadal P, Canela N, Katakis Y, O'Sullivan CK (2011) Extraction, isolation, and characterization of globulin proteins from *Lupinus albus*. J Agric Food Chem 59:2752–2758

Chapter 15

Aptamer-Based Enzyme Capture Assay for Measurement of Plasma Thrombin Levels

Jens Müller, Tobias Becher, Günter Mayer, and Bernd Pötzsch

Abstract

The quantification of circulating thrombin is a valuable tool to accurately assess the activity of the blood coagulation system. Here, we describe the combined application of the thrombin-specific reversible active-site inhibitor argatroban and the DNA-aptamer HD1-22 for conduction of an enzyme capture assay for reliable measurement of plasma thrombin levels.

Key words Thrombin, Plasma, Active-site inhibitor, Argatroban, Aptamer, Enzyme capture assay

---

1 Introduction

Thrombin is a multifunctional enzyme that plays an important role within the process of blood coagulation. Besides the activation of blood platelets and plasmatic coagulation cofactors, one of the main functions of thrombin is the conversion of fibrinogen to fibrin, leading to the formation of a stable clot. To prevent the development of thrombosis, the activity of thrombin is tightly regulated by several endogenous inhibitory mechanisms [1].

The plasma level of active thrombin represents a useful biomarker for the assessment of the activity of a patient's blood coagulation system. To date, the generation of thrombin is determined indirectly by the measurement of plasma thrombin-antithrombin complexes (TAT) or prothrombin activation peptides (F1.2) [2]. Since both markers accumulate in the circulation, however, corresponding concentrations do not reflect the coagulation status at the time point of sample taking.

Günter Mayer (ed.), *Nucleic Acid Aptamers: Selection, Characterization, and Application*, Methods in Molecular Biology, vol. 1380, DOI 10.1007/978-1-4939-3197-2_15, © Springer Science+Business Media New York 2016

In comparison to the quantification of plasma TAT or F1.2, the measurement of circulating levels of active thrombin is far more challenging. In addition to the need for a specific ligand that allows for the discrimination between the enzyme and its structural similar zymogen, a strategy to prevent the fast inhibition of free thrombin ex vivo is required [3].

In this chapter, the combined application of a thrombin-specific DNA-aptamer and a reversible active site inhibitor is described that allows for the quantification of plasma thrombin levels. Besides the basic methodological details, we also go into details about the potential pitfalls that come with this oligonucleotide-based enzyme capture assay.

2 Materials

Prepare all solutions using sterile ultrapure water and analytical grade reagents. Prepare and store all reagents at room temperature (unless indicated otherwise).

2.1 Blood Sampling

1. Blood sampling set: Sterile (winged) blood collection/infusion set with tubing (up to 30 cm) and short (up to 20 mm) cannula with an inner diameter of at least 0.8 mm (21G) but preferable 1.2 mm (18G) (*see* **Note 1**).

2. Basic syringe-like blood collection tubes: e.g., S-Monovette® (Sarstedt, Nürnbrecht, Germany) containing sodium-citrate (10.6 mM final concentration in e.g., 3 ml of taken whole blood) (*see* **Note 2**).

3. 190 mM Argatroban: e.g., Argatra® (Mitsubishi Pharma, Düsseldorf, Germany) (*see* **Note 3**). Store stock solution at 4 °C upon opening of original vial.

4. 154 mM NaCl solution.

2.2 Oligonucleotide-Based Enzyme Capture Assay (OECA)

1. Fluorometer for 96-well plates with appropriate filter set for measurement of 7-amino-4-methyl coumarins (AMC) (Ex/Em = 340–360/440–460 nm) (*see* **Note 4**).

2. Framed Maxisorp Fluoronunc 8 × 12 microtiter modules (white) (Nunc, Thermo Fisher Scientific) (*see* **Note 5**).

3. Adhesive polyester film (Platemax®, Axygen, Union City, CA, USA).

4. Biotin-labeled BSA: Store lyophilized powder at 4 °C until dissolved. Aliquot and store resolved stock solution at ≤−40 °C. Prevent freeze–thaw cycles.

5. Streptavidin (ultrapure) (AppliChem, Darmstadt, Germany): Store lyophilized powder at 4 °C until dissolved. Aliquot and store resolved stock solution at ≤−40 °C. Prevent freeze–thaw cycles.

6. Protease-free bovine serum albumin (BSA): Store at 4 °C.

7. 3′-Biotinylated thrombin-binding DNA-aptamer HD1-22 (5′-GGTTGGTGTGGTTGGAAAAAAA-AAAAAAAAGTCC GTGGTAGGGCAGGTTGGGGTGACT-3′; Page purified) (Microsynth, Balgach, Switzerland) (*see* **Note 6**). Store lyophilized powder at 4 °C until dissolved. Aliquot and store resolved stock solutions (e.g., 100 μM) at ≤ –20 °C until used.

8. Human α-thrombin with defined specific activity (Haematologic Technologies, Essex Junction, VT, USA) (*see* **Note 7**): Store original vial at –20 °C. For long-term storage, dilute material in PBS, pH 7.4, containing 10 mg/ml BSA. Prepare aliquots and store at < –70 °C until use. Prevent freeze–thaw cycles.

9. Normal citrate-anticoagulated human plasma (*see* **Note 8**): Store at < –40 °C until used. Prevent freeze–thaw cycles.

10. Highly pure sterile-filtered dimethyl sulfoxide (DMSO).

11. Sensitive fluorogenic peptide substrate for thrombin detection (Boc-Asp(OBzl)-Pro-Arg-AMC   [I-1560])   (Bachem, Bubendorf, Switzerland) (*see* **Note 9**): Store lyophilized powder at < –15 °C. Dissolve in 10 % DMSO in sterile water: remove powder from the original plastic tube and add into a glass tube. Dissolve powder to reach a concentration of 10 mM. Store aliquots at –20 °C until used. Prevent freeze–thaw cycles.

12. Coating buffer: 30 mmol/L Na_2CO_2, 200 mmol/L $NaHCO_3$, pH 9.0.

13. PBS buffer (general composition): 137 mmol/L NaCl, 2.7 mmol/L KCl, 9.6 mmol/L Na_2HPO_4, and 1.5 mmol/L KH_2PO_4, pH 7.4.

14. Aptamer buffer: PBS, pH 7.4, 3 mmol/L $MgCl_2$, 1 mg/ml BSA.

15. Washing buffer: PBS, pH 7.4, 3 mmol/L $MgCl_2$, 0.05 % Tween 20.

16. Dilution buffer: PBS, pH 7.4, 1 mg/ml BSA; 200 μM Argatroban.

17. Blocking buffer: PBS, pH 7.4, 20 mg/ml BSA, 0.05 % Tween 20.

18. Substrate buffer: 10 mM Tris–HCl, 154 mM NaCl, pH 8.5.

3 Methods

Once formed and released into the circulation, thrombin is promptly inactivated, mainly by complex formation with the serine protease inhibitor (serpin) antithrombin [1]. Accordingly, the half-life of thrombin in plasma has been found to be about 1 min [4]. Since this process of inactivation also takes place ex vivo, a strategy to stabilize the amount of free thrombin during and after the sampling of whole blood is needed [3].

The interaction between thrombin and a serpin involves the enzyme's active site [5]. Thus, blocking of the latter efficiently prevents complex formation. During the assay described here, the thrombin-specific, reversible active site inhibitor argatroban is added to the citrate blood collection tubes. While citrate binds Ca^{2+}-ions and thereby prevents the clotting of taken blood ex vivo, argatroban binds to the active site of thrombin molecules, thereby preventing their inactivation by complex formation [3]. After centrifugation of whole blood, processed plasma is added to the wells of microtiter-modules that have been coated with the thrombin-specific, bivalent DNA-aptamer HD1-22 [6, 7]. Since HD1-22 recognizes two distinct exosites of the thrombin molecule, the presence of argatroban does not interfere with binding. During the subsequent washing step, argatroban is completely removed together with plasma remains, making the active sites accessible for a fluorogenic peptide substrate that is finally used for the quantification of the immobilized thrombin molecules [3].

In this paragraph, all steps needed for the measurement of plasma thrombin levels by the aptamer-based enzyme capture assay are described. Carry out all procedures at room temperature unless otherwise specified. During incubation steps, wells should be generally sealed with adhesive polyester film and stored in the dark. Care should be taken to avoid temperature differences within the microtiter modules. For washing, if not otherwise stated, discard incubated solution and rinse wells three times with 300 µL of washing buffer. An automated plate washer might be used for this purpose (e.g., Biotek ELx50, Biotek, Bad Friedrichshall, Germany). After, washing, tap inverted wells gently on clean adsorbent paper to remove any excess solution.

3.1 Preparation of Thrombin Blood Collection Tubes

1. Add Argatroban to citrate blood collection tubes in order to yield a final concentration of 100 µM in the blood sample: For example, pre-dilute Argatra® 1:50 in 0.9 % NaCl solution (e.g., 490 µl NaCl + 10 µl Argatra®) to reach a concentration of 3.8 mM (see **Note 10**). Subsequently, add 79 µl of this solution to a 3 ml citrate blood collection tube. Store tubes at 2–8 °C for up to 2 months. Bring to RT before use.

3.2 Blood Collection and Handling of Samples

1. Collect blood into prepared thrombin blood collection tubes (see **Note 11**).

2. Store blood samples at RT and centrifuge within 8 h (see **Note 12**).

3. Centrifuge the blood sample at $2500 \times g$ for 15 min.

4. Store processed plasma at ≤ -40 °C for up to 6 months until analyzed (see **Note 12**).

3.3 Preparation of Aptamer-Loaded Microtiter Modules

1. Add 100 μl of coating buffer containing 10 μg/ml of biotin-labeled BSA into designated wells. Incubate overnight at 4 °C.

2. Wash wells and add 100 μL of washing buffer containing 1 mg/ml BSA and 10 μg/ml streptavidin into the wells and incubate for 1 h.

3. Wash wells and block by adding 200 μL/well of blocking buffer. Incubate for 2 h.

4. Aspirate/remove the blocking buffer and tap inverted wells gently on clean adsorbent paper to remove excess solution.

5. Load 3′-biotinylated aptamers into the primed wells: dilute 3′-biotinylated HD1-22 to a concentration of 10 nmol/L using aptamer buffer and add 100 μl of this solution to each well. Incubate for 1 h.

6. Wash wells and store sealed framed aptamer-loaded modules at 4 °C for up to 2 month until use.

3.4 Preparation of Plasma-Based Thrombin Calibrators

1. Prepare Argatroban-primed (200 μM final concentration) normal citrate-anticoagulated human plasma: For the preparation of 2000 μl, pre-dilute Argatra® 1:50 in 0.9 % NaCl solution (e.g., 490 μl NaCl + 10 μl Argatra®) to reach a concentration of 3.8 mM (*see* **Note 10**). Subsequently, add 105 μl of this solution to 1895 μl of citrate-anticoagulated human plasma and mix well.

2. For preparation of plasma thrombin calibrators, perform ½-log dilution series of human α-thrombin in Argatroban-primed plasma, starting from 40 mU/ml (*see* **Note 7**). Prepare 250 μl of each dilution, covering a concentration range down to 0.13 U/ml. Apply remaining Argatroban-primed plasma as the blank calibrator.

3.5 Oligonucleotide-Based Enzyme Capture Assay (OECA)

1. Frame needed number of aptamer-loaded microtiter modules and bring to RT.

2. Thaw frozen calibrators and samples in a water bath at 37 °C. Subsequently, bring to RT and assay as described below. Samples and calibrators are tested undiluted and should be run in duplicate.

3. Pipette 100 μl of calibrators and samples into designated wells. Incubate for 1 h.

4. Prepare 1 ml of substrate solution for each 8 wells right in time before the end of the incubation. Dilute the stock solution of the fluorogenic substrate (30 mM) 1 in 100 using substrate buffer to yield a final substrate concentration of 300 μM. Store at RT in the dark until use.

5. Empty the wells using an eight-channel multi-pipette. Take care that fresh tips for each module (strip) are used to avoid

cross-contamination. Subsequently, manually add 250 µl of washing buffer to each well and wash additional three times with washing buffer (300 µl/well) (*see* **Note 13**).

6. Pipette 100 µl of the freshly prepared substrate solution to each well. Measure fluorescence at 360[ex]/460[em] nm within 1–2 min.

7. Seal plate and incubate for 2 h.

8. Measure fluorescence at 360[ex]/460[em]. Calculate the mean net change in fluorescence (dFU) for each calibrator.

9. Construct a standard curve by plotting the dFU observed for each calibrator versus the corresponding concentration of thrombin in mU/ml. Use an appropriate curve fit algorithm (e.g., 4-parameter logistic function) to calculate the amount of thrombin present in the plasma samples (Fig. 1a). Calculate mean values from duplicate samples after conversion of dFU to corresponding plasma thrombin concentrations (Fig. 1b). Ensure that quantitative values are above the lower limit of quantification (LLOQ) of the assay. Thrombin plasma levels in healthy individuals are typically below the LOD (*see* **Note 14**). Consider any medications or conditions that may directly or indirectly (artificially) influence assay results (*see* **Note 15**).

10. Samples which yield values above the highest standard must be pre-diluted with dilution buffer and retested. Multiply the results by the dilution factor in order to obtain the concentration of thrombin in the original sample (*see* **Note 16**).

4 Notes

1. We found that the use of cannula with an inner diameter of less than 0.8 mm increases the risk of artificial generation of thrombin during the course of blood collection (*see* **Note 11**).

2. Use syringe-based (Monovette) rather than vacuum-based (Vacutainer) blood collection tubes to allow for convenient addition of Argatroban solution (*see* **Note 10**). Furthermore, higher shear rates which result from the use of vacuum tubes may also facilitate artificial thrombin generation during blood sampling.

3. We use medical-grade Argatroban (Argatra®) for all experiments. Consider that Argatra® is a prescription drug. In case of legal difficulties, Argatroban might also be purchased as a research reagent. However, we do not have any experience with the use of such preparations.

4. Use appropriate filter sets for the fluorophore AMC (*see* **Note 9**) and assess optimal gain settings. For the thrombin assay described here, white microtiter modules should be used (*see*

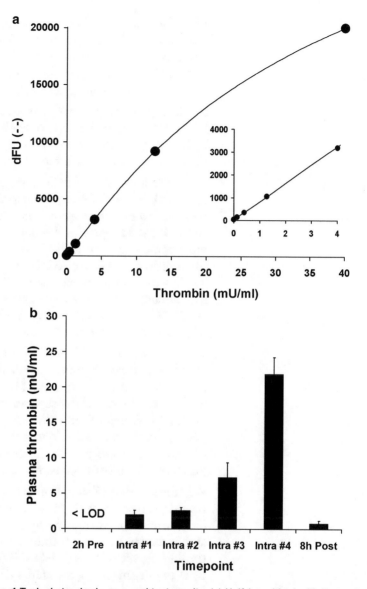

Fig. 1 Typical standard curve and test results. (**a**) Half-logarithmic dilution series of human alpha-thrombin in argatroban-primed plasma. Shown interpolation was done by 4-parameter-logistic curve fit. *Inlet* shows the lower concentration range. (**b**) Typical results as measured during hip replacement surgery. While levels before surgery were found to be below the LOD of the assay, plasma thrombin concentrations continuously increased during surgery and rapidly decreased afterwards

Note 5). Accordingly, gain settings of the fluorometer must be adjusted accordingly in order to prevent saturation of signals.

5. The AMC-based fluorogenic substrate (*see* **Note 9**) leads to significant background fluorescence of the solution. Thus, the general benefit of very-low-background fluorescence of black

microtiter modules becomes secondary. In fact, the white modules, which are applicable for both fluorescence and luminescence measurements, yielded better performance with respect to assay sensitivity and reproducibility.

6. The aptamer HD1-22 must be ordered with a 3′-biotin tag. It has been previously shown that the modification of the 5′-end of the HD1 aptamer sequence leads to reduced binding affinity [8]. Page purification should be chosen to ensure best possible purity of full-length molecules.

7. The application of well-defined preparations of highly purified human alpha-thrombin for preparation of thrombin standards is critical to ensure reproducible results between batches and runs. Only order purified human alpha thrombin. Each batch must come with a certificate that states the specific activity of the preparation in terms of (NIH) units/mg [9]. Do not use batches that show an activity of < 3500 units/mg. Thrombin may be supplied in 50 % (vol/vol) glycerol/H_2O (to be stored at –20 °C). However, we observed highest long-term stability when storing appropriately diluted material at ≤ –70 °C.

8. We use in house preparations of citrate-anticoagulated plasma: blood from healthy donors is taken into donation bags containing sodium-citrate solution (10 mM final concentration in taken whole blood). After separation of cells by centrifugation, processed plasma of individual donations is mixed to achieve pooled normal plasma. Aliquots are stored at –40 °C until use. Alternatively, (pooled) citrate-anticoagulated plasma may be purchased from commercial sources.

9. We found that 7-amino-4-methylcoumarin (AMC)-based peptide substrates with high conversion rates show the best compromise between assay sensitivity and reproducibility. As an alternative to the substrate listed in Subheading 2.2, also the fluorogenic substrate H-D-CHA-Ala-Arg-AMC, as offered by Pentapharm (Basel, Switzerland) can be used. Due to inadequate sensitivity, the use of chromogenic substrates is not recommended. In comparison to AMC, the use of a rhodamine-110 (Rh110)-based peptide substrate may improve overall assay sensitivity. However, we observed an unsatisfactory reproducibility of results at low concentrations when using an Rh-110-based substrate for quantification of plasma thrombin levels.

10. Medical-grade Argatroban (Argatra®) is supplied by Mitsubishi Pharma at a concentration of 100 mg/ml (190 mM) in ethanol-based solution. Due to its limited solubility in water, Argatroban will precipitate during the described first dilution step. In order to ensure complete dissolving, the following measures should be taken: Use a 100 μl pipette tip to take up

10 µl of the original Argatra® solution. Wipe off any droplets remaining on the outside of the tip and pipette the 10 µl quickly into 490 µl of 0.9 % NaCl solution without coming into contact with the reaction tube. Precipitates will be clearly visible. Directly close and vortex the reaction tube to completely redissolve the aragtroban. Inspect the used pipette tip and reaction tube: discard and repeat described procedure in case of any remaining precipitates. If problems remain, warm both Argatra® and the NaCl solution to 37 °C before mixing them together as described above.

11. Please consider that prepared thrombin blood collection tubes might not be sterile after addition of the argatroban-solution and should therefore not be used for blood collection via direct venipuncture/phlebotomy. Blood should be collected using a sterile collection/infusion set with tubing. Also keep in mind that the process of venipuncture represents a vascular lesion itself and therefore bears the risk of artificial thrombin generation that may interfere with the measurement of circulating thrombin levels. Thus, in order to minimize such effects, the following measures should be taken: the first blood sample (up to 3 ml) should not be used for thrombin determination, therefore use other blood collection tube, e.g. EDTA-tube, for other analysis than thrombin determination first and then immediately draw blood into the thrombin blood collection tube(s). Discard material in case of initially unsuccessful puncture attempt or obviously impaired blood flow during sampling. Do not puncture the same vein again but preferably change to the contralateral limb for second attempt.

12. Ensure storage of filled thrombin blood collection tube at RT. We observed a drop in thrombin activity to 90 % in case tubes were cooled before centrifugation [3]. However, processed plasma should be stored on ice or frozen at ≤ -40 °C until analyzed.

13. Due to the high sensitivity of the assay, there is a risk of cross-contamination between wells when using an automated plate washer to aspirate the samples from the wells. Thus, original samples should be manually removed from the wells using disposable tips. After subsequent manual addition of washing buffer, the automated washing procedure might be applied.

14. Lab specific validation of the assay's key characteristics should be performed. Besides reproducibility, also the LLOQ and the limit of detection (LOD) of the assay should be determined [3, 10]. Argatroban-primed plasma should be used for the preparation of corresponding thrombin dilutions [3]. In our lab, the following values were determined: LLOQ: 0.039 ng/ml (1.08 pM). LOD: 0.017 ng/ml (0.47 pM). According to

the specific activity of the used thrombin preparation, these values correspond to 0.15 mU/ml (LLOQ) and 0.06 mU/ml (LOD). When plasma samples obtained from 20 healthy blood donors were studied, levels of free thrombin were found to below the LLOQ ($n=5$) or the LOD ($n=15$) of the assay. In contrast, highly increased plasma thrombin levels were found during the course of hip replacement surgery [3].

15. The DNA aptamer HD1-22 simultaneously binds to exosites I and II of thrombin [6, 7]. Since exosite II is the heparin-binding site, the presence of heparins might lead to artificially decreased levels of plasma thrombin due to impaired binding of the heparin-thrombin complex. The same is true for direct thrombin inhibitors that simultaneously target exosite I and the active site (e.g., bivalirudin). In contrast, the presence of active site inhibitors like argatroban or dabigatran [11] may lead to artificially increased levels due to impaired complex-formation between thrombin and the serpins [5]. However, since all directly or indirectly acting anticoagulants lead to down-regulation of thrombin generation in vivo, such effects may be secondary due to resulting non-detectable plasma thrombin levels.

16. Since the activity of free thrombin is highly regulated in vivo, observed plasma thrombin levels are normally found within the 40 mU/ml range as covered by the standard curve. Especially within the lower concentration range, plasma (matrix) calibrators are needed to correct for enzymatic plasma background activity, which trigger unspecific fluorescence generation [3, 10]. In case the pre-dilution of samples due to high plasma thrombin levels is needed; however, this aspect becomes secondary, allowing the use of buffer instead of plasma for sample dilution.

References

1. Mann KG (2003) Thrombin formation. Chest 124(3 Suppl):4S–10S

2. Bailey MA, Griffin KJ, Sohrabi S, Whalley DJ, Johnson AB, Baxter PD, Ariëns RA, Scott DJ (2013) Plasma thrombin-antithrombin complex, prothrombin fragments 1 and 2, and D-dimer levels are elevated after endovascular but not open repair of infrarenal abdominal aortic aneurysm. J Vasc Surg 57:1512–1518

3. Müller J, Becher T, Braunstein J, Berdel P, Gravius S, Rohrbach F, Oldenburg J, Mayer G, Pötzsch B (2012) Profiling of active thrombin in human blood by supramolecular complexes. Angew Chem Int Ed 50:6075–6078

4. Rühl H, Müller J, Harbrecht U, Fimmers R, Oldenburg J, Mayer G, Pötzsch B (2012) Thrombin inhibition profiles in healthy indi-

viduals and thrombophilic patients. Thromb Haemost 107:848–853

5. Huntington JA (2013) Thrombin inhibition by the serpins. J Thromb Haemost 11(Suppl 1):254–264

6. Müller J, Wulffen B, Pötzsch B, Mayer G (2007) Multidomain targeting generates a high-affinity thrombin-inhibiting bivalent aptamer. Chembiochem 8:2223–2226

7. Müller J, Freitag D, Mayer G, Pötzsch B (2008) Anticoagulant characteristics of HD1-22, a bivalent aptamer that specifically inhibits thrombin and prothrombinase. J Thromb Haemost 6:2105–2112

8. Buff MC, Schäfer F, Wulffen B, Müller J, Pötzsch B, Heckel A, Mayer G (2010)

Dependence of aptamer activity on opposed terminal extensions: improvement of light-regulation efficiency. Nucleic Acids Res 38:2111–2118

9. Whitton C, Sands D, Lee T, Chang A, Longstaff C (2005) A reunification of the US ("NIH") and International Unit into a single standard for Thrombin. Thromb Haemost 93:261–266

10. Müller J, Friedrich M, Becher T, Braunstein J, Kupper T, Berdel P, Gravius S, Rohrbach F, Oldenburg J, Mayer G, Pötzsch B (2012) Monitoring of plasma levels of activated protein C using a clinically applicable oligonucleotide-based enzyme capture assay. J Thromb Haemost 10:390–398

11. Harbrecht U (2011) Old and new anticoagulants. Hamostaseologie 31:21–27

Chapter 16

Application of Aptamers in Histopathology

Sarah Shigdar, Li Lv, Lifen Wang, and Wei Duan

Abstract

Aptamers are proving to be exceedingly effective in a number of applications. Given the disadvantages of antibodies, such as batch-to-batch variation and cross-reactivity, aptamers have the potential to revolution-ize the field of histopathology due to their high specificity and the ease of their synthesis and modification. Here, we describe a chromogenic staining method for paraffin-embedded tissue sections with FITC-labeled aptamers.

Key words Aptamers, Chemical antibodies, Chromogenic staining, Diagnostics, Histology, Immunohistochemistry, Paraffin-embedded tissues, Pathology

1 Introduction

Monoclonal antibodies have helped to revolutionize the field of diagnostics, especially in pathology where the diagnosis of cellular markers can aid in diagnosis and guide therapeutic treatments for a number of cancers. However, antibodies are not the perfect reagent for laboratory diagnostics. Two recent articles have highlighted the very real technical challenges facing any scientist wishing to use antibodies as prognostic or predictive biomarkers [1, 2]. The most likely reason why, even with all these pitfalls, such as batch-to-batch variation and a lack of validation, antibodies have been so extensively utilized is the lack of a "better" alternative. However, the discovery of a technique in 1990 that is able to produce "chemical antibodies" has now put another class of reagents at the hands of scientists and pathologists. These short single stranded DNA or RNA molecules, otherwise known as aptamers, are produced via an in vitro process known as the systematic evolution of ligands by exponential enrichment, or SELEX [3, 4]. This process involves the iterative incubation of a randomized library of aptamers with the target of choice and generates highly specific and sensitive aptamers. These aptamers can then be easily modified with different reporter molecules to suit a particular application. Indeed,

Günter Mayer (ed.), *Nucleic Acid Aptamers: Selection, Characterization, and Application*, Methods in Molecular Biology, vol. 1380, DOI 10.1007/978-1-4939-3197-2_16, © Springer Science+Business Media New York 2016

aptamers can be easily conjugated to fluorescent molecules for fluorescence microscopy [5, 6], or to biotin for standard immuno-histochemistry protocols [7]. Several recent reports have demonstrated that aptamers are more sensitive than antibodies and in a timelier manner [7–9], though as with antibody immunohisto-chemistry (Ab-IHC), aptamer immunohistochemistry (Apt-IHC) requires some optimization. One of the main issues is the nonspe-cific attachment of aptamers to the nuclei of cells in both frozen and paraffin-embedded tissues (PET), and it has been suggested that this is due to electrostatic attraction of the polyanion aptamers to positively charged sites in the nuclei [8]. Indeed, the use of dex-tran sulphate and heparin have long been used for in situ hybrid-ization as a means of reducing background and accelerating the rate of nucleic acid hybridization [10–12]. When included in the binding buffer, these two components allowed successful staining of a number of PET sections with several RNA aptamers [9] and we have tested this protocol with RNA and DNA aptamers tar-geted against other cell surface markers [6, 13] with success as well (unpublished data). We have also found that using the aptamers at concentrations slightly above the equilibrium dissociation constant (KD) of the aptamer to work well, though optimization of the concentration with positive and negative control slides is required. Additional optimisation of the incubation time with the aptamer is also required. We have found that, especially with the aptamer directed against EpCAM, an incubation time of 15 min at 37 °C to be adequate for superior staining quality. An incubation time of less or more than 15 min resulted in inferior staining. As with Ab-IHC, all steps in the protocol for Apt-IHC need to be optimized, though the authors hope that the protocol presented here will aid researchers wanting to test their own or published aptamers for Apt-IHC.

2   Materials (*See* Note 1)

2.1 Antigen Retrieval Buffers

1. Tris–EDTA buffer: 10 mM Tris base, pH 9.0, 1 mM EDTA solution, 0.05 % Tween 20. Weigh 1.21 g of Tris base and 0.37 g of EDTA and transfer to a beaker. Add approximately 800 mL of distilled water to the beaker and check the pH of the buffer. The pH is usually at 9.0. Add 0.5 mL of Tween 20 and add the remaining distilled water to make a solution of 1000 mL. Mix well and store this solution in a glass bottle at room temperature for 3 months or 4 °C for longer storage.

2. Sodium citrate buffer: 10 mM Sodium citrate, pH 6.0, 0.05 % Tween 20. Weigh 2.94 g of trisodium citrate (dehydrate) and transfer to a beaker. Add approximately 800 mL of distilled water to the beaker and adjust the pH to 6.0 with 1 N HCl.

Add 0.5 mL of Tween 20 and add the remaining distilled water to make a solution of 1000 mL. Mix well and store this solution in a glass bottle at room temperature for 3 months or 4 °C for longer storage.

2.2 Antibodies and Commercial Reagents

1. Anti-FITC HRP antibody (AB6656, Abcam).

2. DAB+ Chromogen staining solution.

3. Bovine serum albumin, tRNA from baker's yeast, and goat serum.

3 Methods

Carry out all procedures at room temperature unless otherwise stated. Protocols were optimized using 5 μm sections.

1. Deparaffinize tissue sections using histoclear for 5 min twice and rehydrate using serial dilutions of alcohol (100 %, 95 %, 70 %, for 5 min each).

2. Store tissue sections in distilled water while the antigen retrieval buffer is heated.

3. Heat the antigen retrieval buffer (*see* **Note 2**) in the microwave to achieve a temperature of 95–98 °C.

4. Place slides in antigen retrieval buffer in a sealed container and microwave on low power for 5–20 min (*see* **Note 3**).

5. Once the slides have been microwaved for a sufficient amount of time, take the lid off the container and top up the solution with fresh antigen retrieval buffer (at room temperature). Allow to cool on the bench for approximately 20–30 min.

6. Wash the slides in PBS and block using 3 % hydrogen peroxide for 15 min (*see* **Note 4**).

7. Wash the slides in PBS again and prepare a horizontal staining chamber (*see* Fig. 1 and **Note 5**) for the blocking step.

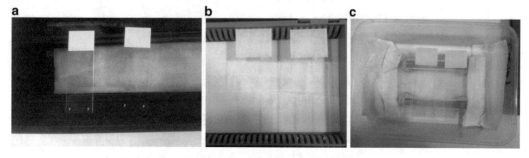

Fig. 1 Examples of horizontal staining chambers. (**a**) A hybridization slide box; (**b**) a slide box; (**c**) a plastic container using small 5 mL tubes to keep the slides raised to allow the placement of wet tissues or paper towel to maintain a moist environment during the incubation steps

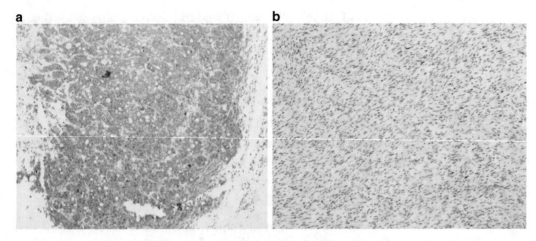

Fig. 2 Indicative staining of paraffin-embedded tumors using the FITC-conjugated anti-EpCAM aptamer Ep23 with the DAB chromogenic stain: (**a**) EpCAM positive breast tumor; EpCAM negative glioblastoma tumor. Magnification: ×40

8. Prepare the blocking reagents—goat serum at 10 %, BSA at 1 mg/mL, and tRNA at 0.1 mg/mL.

9. Add 150 μL of blocking reagent per slide and cover with parafilm. Incubate for 30 min to 1 h.

10. During the blocking step, prepare the FITC conjugated aptamer (*see* **Note 6**).

11. Following incubation with blocking reagent, wash the slides with PBS three times for 5 min each.

12. Add 100–200 μL aptamer to the slides, cover with parafilm, and incubate the slides in a horizontal chamber (*see* **Note 7**).

13. Wash the slides with PBS three times for 5 min each.

14. Add 100–200 μL anti-FITC HRP antibody to the slides, cover with parafilm, and incubate the slides in a horizontal chamber for 2 h.

15. Wash the slides with PBS three times for 5 min each.

16. Incubate the slides with 3,3′-diaminobenzidine (DAB) for 10 min (*see* **Note 8**).

17. Wash the slides in distilled water for 5 min.

18. Counterstain the slides with hematoxylin (*see* **Note 9**).

19. Dehydrate the slides in alcohol (95 % and 100 %).

20. Process the slides through histoclear and coverslip using DPX (Distrene, Plasticiser, Xylene) mounting media.

21. Visualize results using a light microscope (Fig. 2).

4   Notes

1. Prepare all solutions using ultrapure water and analytical grade reagents. Prepare and store all reagents at room temperature (unless otherwise stated). Concentrations of aptamers and antibodies need to be optimised by the end user. Aptamers are diluted in sterile PBS containing the correct salt concentrations used in initial selection experiments. Antibodies are diluted in sterile PBS.

2. Not all antigens will require antigen retrieval. However, the detection of many antigens can be significantly improved using antigen retrieval. Antigen retrieval will depend on the antigen in question. Not all antigens will require heat-induced antigen retrieval, and there are a number of proteolytic-induced epitope retrieval methods that can be employed. The authors direct readers to the website "IHC World" for a full listing of up-to-date protocols (http://www.ihcworld.com/epitope_retrieval.htm).

3. Coated slides, such as poly-l-lysine-coated slides, should be used to prevent tissues sections floating off the slide during heat-induced antigen retrieval. Additionally, low power should be used for antigen retrieval to prevent the tissue sections detaching from the slide due to bubble formation if the microwave overheats the solution.

4. Some cells contain endogenous peroxidase and can result in high, non-specific background staining if a horseradish peroxidase conjugated antibody is used. Quenching of endogenous peroxidase can be accomplished using hydrogen peroxide and this can be completed at different points in the protocol: e.g., prior to antigen retrieval; prior to aptamer incubation; following aptamer incubation; or following secondary antibody incubation. The authors prefer to complete the endogenous peroxidase quenching step following antigen retrieval. In addition, some cell surface markers are sensitive to hydrogen peroxidase and can reduce staining intensity and so quenching should be completed following primary or secondary aptamer/antibody incubation. Hydrogen peroxide can be diluted with methanol, PBS, or distilled water. Some antigens can be highly sensitive to methanol/hydrogen peroxide quenching, so the diluent of choice for cell surface or membrane markers is PBS/hydrogen peroxide.

5. A horizontal staining chamber can be created using a tip box lid or a plastic slide box. The slides need to be lifted off the base of the container as wet tissue or paper towel should be placed in the base to keep the slides moist during the incubation step.

6. The concentration of aptamer will need to be optimised for immunohistochemical staining. It is prudent to start with aptamer concentration ranging from 50 to 400 nM. Users must check the salt constituents (especially $MgCl_2$) of the buffer used for aptamer selection to ensure the optimal salt concentrations are used in the aptamer folding buffer. Additionally, the aptamer may require the addition of dextran sulphate (10 %) and heparin (500 µg/mL) following folding of the aptamer to reduce nonspecific staining.

7. Incubation times will vary, depending on the aptamer. Some aptamers will require a 15 min incubation time at 37 °C, while others may require longer. Suggested incubation times are 10–60 min. Aptamers typically do not require an extended incubation time.

8. The authors have favoured the DAB chromogen solution for visualisation of antigens. However, there are now numerous chromogenic systems available which may offer advantages over DAB, depending on the end-user requirements.

9. Tissue sections are countered stained in hematoxylin using standard protocols to visualise the nuclei of cells.

References

1. Bordeaux J, Welsh A, Agarwal S, Killiam E, Baquero M, Hanna J et al (2010) Antibody validation. Biotechniques 48:197–209

2. Perkel JM (2014) The antibody challenge. Biotechniques 56:111–114

3. Ellington AD, Szostak JW (1990) In vitro selection of RNA molecules that bind specific ligands. Nature 346:818–822

4. Tuerk C, Gold L (1990) Systematic evolution of ligands by exponential enrichment: RNA ligands to bacteriophage T4 DNA polymerase. Science 249:505–510

5. Shigdar S, Lin J, Yu Y, Pastuovic M, Wei M, Duan W (2011) RNA aptamer against a cancer stem cell marker epithelial cell adhesion molecule. Cancer Sci 102:991–998

6. Shigdar S, Qiao L, Zhou SF, Xiang D, Wang T, Li Y et al (2013) RNA aptamers targeting cancer stem cell marker CD133. Cancer Lett 330:84–95

7. Zeng Z, Zhang P, Zhao N, Sheehan AM, Tung CH, Chang CC et al (2010) Using oligonucleotide aptamer probes for immunostaining of formalin-fixed and paraffin-embedded tissues. Mod Pathol 23:1553–1558

8. Gupta S, Thirstrup D, Jarvis TC, Schneider DJ, Wilcox SK, Carter J et al (2011) Rapid histochemistry using slow off-rate modified aptamers with anionic competition. Appl Immunohistochem Mol Morphol 19: 273–278

9. Shigdar S, Qian C, Lv L, Pu C, Li Y, Li L et al (2013) The use of sensitive chemical antibodies for diagnosis: detection of low levels of EpCAM in breast cancer. PLoS One 8, e57613

10. Wahl GM, Stern M, Stark GR (1979) Efficient transfer of large DNA fragments from agarose gels to diazobenzyloxymethyl-paper and rapid hybridization by using dextran sulfate. Proc Natl Acad Sci U S A 76:3683–3687

11. van Gijlswijk RP, Wiegant J, Raap AK, Tanke HJ (1996) Improved localization of fluorescent tyramides for fluorescence in situ hybridization using dextran sulfate and polyvinyl alcohol. J Histochem Cytochem 44:389–392

12. Singh L, Jones KW (1984) The use of heparin as a simple cost-effective means of controlling background in nucleic acid hybridization procedures. Nucleic Acids Res 12:5627–5638

13. Song Y, Zhu Z, An Y, Zhang W, Zhang H, Liu D et al (2013) Selection of DNA aptamers against epithelial cell adhesion molecule for cancer cell imaging and circulating tumor cell capture. Anal Chem 85:4141–4149

Chapter 17

Aptamer Stainings for Super-resolution Microscopy

Maria Angela Gomes de Castro, Burkhard Rammner, and Felipe Opazo

Abstract

Fluorescence microscopy is an invaluable tool to visualize molecules in their biological context with ease and flexibility. However, studies using conventional light microscopy have been limited to the resolution that light diffraction allows (i.e., ~200 nm). This limitation has been recently circumvented by several types of advanced fluorescence microscopy techniques, which have achieved resolutions of up to ~10 nm. The resulting enhanced imaging precision has helped to find important cellular details that were not visible using diffraction-limited instruments. However, it has also revealed that conventional stainings using large affinity tags, such as antibodies, are not accurate enough for these imaging techniques. Since aptamers are substantially smaller than antibodies, they could provide a real advantage in super-resolution imaging. Here we compare the live staining of transferrin receptors (TfnR) obtained with different fluorescently labeled affinity probes: aptamers, specific monoclonal antibodies, or the natural receptor ligand transferrin. We observed negligible differences between these staining strategies when imaging is performed with conventional light microscopy (i.e., laser scanning confocal microscopy). However, a clear superiority of the aptamer tag over antibodies became apparent in super-resolved images obtained with stimulated emission depletion (STED) microscopy.

Key words Aptamers, Antibodies, Microscopy, Super-resolution, STED, Affinity probes, Staining

1 Introduction

Fluorescence microscopy allows the localization and tracking of fluorescent elements embedded in complex biological samples. This technique has gained much popularity due to its relative simplicity and flexibility. However, fluorescence light microscopy has also been limited to a maximal lateral resolution of 200–300 nm, a boundary imposed by the diffractive nature of light, which was already described by Ernst Abbe in 1873 [1]. This limitation has been circumvented in 1994 [2], giving rise to the field of

Electronic supplementary material: The online version of this chapter (doi:10.1007/978-1-4939-3197-2_17) contains supplementary material, which is available to authorized users.

Günter Mayer (ed.), *Nucleic Acid Aptamers: Selection, Characterization, and Application*, Methods in Molecular Biology, vol. 1380, DOI 10.1007/978-1-4939-3197-2_17, © Springer Science+Business Media New York 2016

super-resolution microscopy. This field advanced rapidly and today, ~20 years later, several different techniques provide images that are not limited by the diffraction of light. Currently, the main super-resolution fluorescence microscopy techniques are *stimulated emission depletion* (STED) [2], *photo-activated localization microscopy* (PALM) [3], *stochastic optical reconstruction microscopy* (STORM) [4], *structural illumination microscopy* (SIM) [5], and *ground-state depletion microscopy followed by individual molecule return* (GSDIM) [6]. While each of these super-resolution techniques has advantages and disadvantages [7, 8], all of them provide researchers with options to pursue questions that would not have been possible to answer using conventional diffraction-limited microscopy [9–12].

For all of these techniques the elements of interest (protein, lipids, sugars) must be linked to fluorescent markers. Typically, this has been achieved using antibodies as affinity tags. In conventional immunostainings the protein of interest is detected by a primary antibody, which is then revealed by a fluorescently labeled secondary antibody. The primary/secondary antibody complex can have a linear length of ~20 nm [13] (Fig. 1a). The large size of antibodies tends to complicate their penetration into tissue, thereby reducing the number of epitopes detected [14, 15]. In addition, the primary-secondary antibody detection system places the reporting fluorophore relatively far from the target molecule, which again lowers the accuracy of the staining. Finally, the fact that both the primary and secondary antibodies used in conventional immunostaining are bivalent (contain two epitope-binding pockets) confers them a tendency to form clusters.

Recently, it has been noted that this imprecision caused by antibodies normally is not relevant in diffraction limited techniques; however, antibodies could introduce a significant amount of inaccuracy in super-resolution imaging [14–16]. Here we describe a staining methodology that uses small aptamers as affinity tags for staining biological samples imaged with confocal and super-resolution STED microscopy. The main advantages of aptamers in the field of advanced microscopy lie in their monomeric binding and their small size, which can be a tenth of a full immunoglobulin (Fig. 1a). The methodology described here is based on a previous study that employed a fluorescently labeled aptamer able to recognize the extracellular domain of the transferrin receptor (TfnR) [15, 17]. The staining was performed in living human A431 cells for several minutes. TfnRs recycle constantly, which resulted in the staining of the network of endosomes that contain these receptors [18]. Contrary to antibodies, aptamer staining resembled the subcellular structures labeled with the natural ligand transferrin used as a fluorescent affinity probe (Fig. 2).

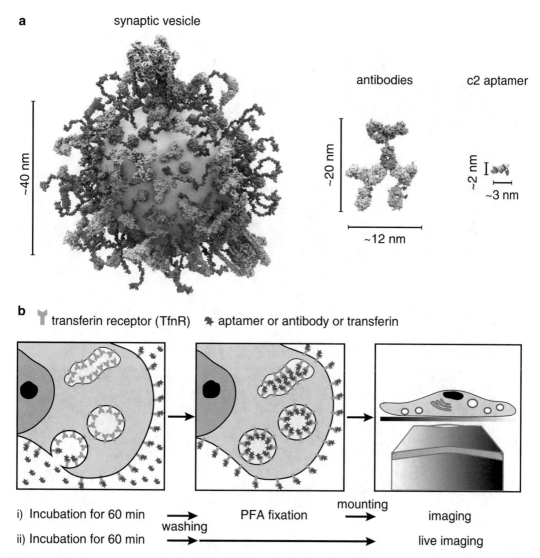

Fig. 1 Probe size and live staining strategy. (**a**) Cartoon comparing a small organelle (synaptic vesicle), an antibody package (one primary bound to two secondary antibodies), and the c2 aptamer. The antibodies were modeled using real structural information (accession number 1hzh from the protein data bank, PDB), the aptamer was modeled after its structure was predicted using the mfold web server. The structural information of proteins in the synaptic vesicle were obtained from PDB and modeled after the work of Wilhelm et al. [20]. Note that aptamers will have a major advantage finding epitopes in a crowded biological environment like a synaptic vesicle surface or lumen. They will position the fluorescent molecule closer to the intended target than antibodies. The *red molecules* coupled to the secondary antibodies and aptamer represent Atto647N fluorophores. (**b**) Simplified schemes of the staining methodology. The staining was performed on living cells for 60 min, to stain most of the endosomal trafficking pathway. After a thorough washing step cells were fixed with paraformaldehyde and imaged (i), or were imaged alive (ii). The rapid and continuous cycling of TfnRs allows the staining of a large fraction of them during the 60 min of incubation with affinity tags

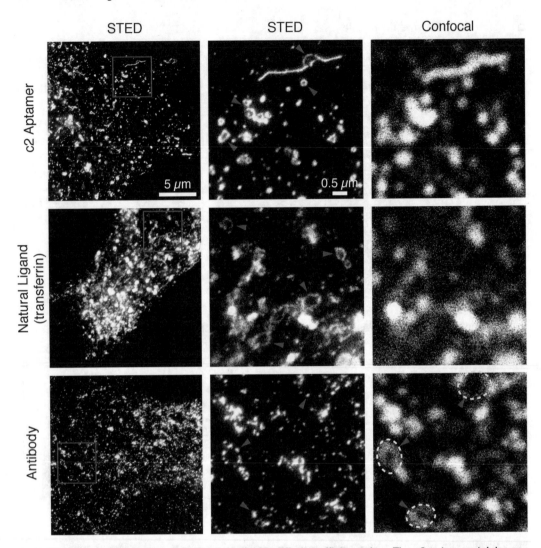

Fig. 2 STED and confocal images of cells stained with different affinity probes. The c2 aptamer staining recognizes round endosomes just as well as the natural ligand transferrin (indicated by *red arrowheads*). Staining using full antibodies resulted in discontinuous stainings of the organelle contours (indicated by *yellow doted ovals* in the confocal image). Note that the difference between the stainings is only evident at diffraction-unlimited resolution (STED). *Red squares* in the *leftmost* column indicate the enlarged area displayed in the *middle* and *right* columns

2   Materials (*See Note 1*)

2.1   Dye Coupling to Aptamers

1. Thiolated minimized c2 aptamer [15, 17] (*see* **Note 2**).

2. Maleimide Atto647N dye (ATTO-TEC GmbH, Siegen, Germany).

3. 1 M Tris (2-carboxyethyl) phosphine hydrochloride stock solution (TCEP).

4. 1 M Triethylammonium acetate buffer pH 7.0 stock solution (TEAA).

5. Biospin 6 column (Bio-Rad, Hercules, USA).

6. Anhydrous dimethyl sulfoxide (DMSO, Invitrogen, Carlsbad, USA).

2.2 Preparation of Glass Cover Slips

1. 1 M HCl solution: Mix 8.3 ml of 37 % HCl with water until reaching 100 ml. Prepare and store at room temperature inside a fume hood.

2. 1 M NaOH solution: Add 4 g of NaOH to 100 ml of water. Prepare and store at room temperature inside a fume hood.

3. pH-meter, Knick Portamess (Knick GmbH & Co, Berlin, Germany).

4. 18 mm diameter round-shaped glass cover slips N°0 (Gerhard Menzel GmbH, Braunschweig, Germany).

5. Poly-L-lysine (PLL) stock solution: Prepare 2 mg/ml stock solution in water and sterilize through syringe filter of 0.2 μm (Minisart® high flow syringe filters, Sartorius, Göttingen, Germany). Make aliquots and store at –20 °C.

6. Cell culture 12-well plates (Thermo Scientific, MA, USA).

2.3 Cell Culture

1. High-glucose (4.5 g/L) Dulbecco's modified Eagle's medium—DMEM (Invitrogen, Carlsbad, USA). Store at 4 °C.

2. Fetal calf serum (FCS): Store at –20 °C (*see* **Note 3**).

3. 10,000 U/ml Penicillin/streptomycin each: Store aliquoted at –20 °C.

4. 200 mM L-Glutamine: Store aliquoted at –20 °C.

5. Trypsin-EDTA solution (0.5 g/L and 0.2 g/L of trypsin and EDTA, respectively).

6. Cell culture dishes of Ø 145×20 mm, 143 cm^2 Cellstar® (Greiner bio-one, Kremsmünster, Austria) (*see* **Note 4**).

7. Complete DMEM medium: DMEM, 15 % FCS, 4 mM l-glutamine, 100 U/ml each penicillin and streptomycin. Store at 4 °C.

8. Imaging DMEM medium: DMEM high glucose, w/o phenol red, 15 % FCS, 4 mM L-glutamine, 100 U/ml each penicillin and streptomycin, 50 mM HEPES buffer, pH 7.4. Sterilized through a 0.2 μm filter system (vacuum system filter, 500 ml capacity with 0.2 μm PES membrane).

9. 1× of commercial Dulbecco's phosphate-buffered saline (DPBS) or 1× PBS (for 1 L of 10× PBS stock solution, dissolve 14.4 g Na$_2$HPO$_4$, 2.4 g KH$_2$PO$_4$, 2 g KCl, and 80 g NaCl in DNase- and RNase-free distilled water (*see* **Note 5**)). Adjust the pH to 7.4. Sterilize by autoclaving. Store at room temperature.

2.4 Aptamer Folding

1. Fluorescently labeled c2 aptamer (30 μM stock solution): Store at −20 °C.

2. Ultrapure DNase- and RNase-free distilled water (Carl Roth GmbH & Co. KG, Karlsruhe, Germany) (*see* **Note 6**).

3. 5× phosphate buffer saline (5× PBS): Dilute the 10× concentrated PBS stock solution in ultrapure DNase- and RNase-free distilled water to avoid particles and contamination filter the solutions inside a cell culture hood with a 0.2 μm syringe filter. Store at room temperature.

4. 25 mM MgCl₂ solution (5× MgCl₂): Dilute the 1 M MgCl₂ stock solution (for 100 ml of 1 M MgCl₂, dissolve 20.3 g MgCl₂ × 6H₂O in DNase- and RNase-free water. Store at room temperature) to 25 mM in DNase- and RNase-free distilled water. To avoid particles and contamination, filter the solutions inside a cell culture hood with a 0.2 μm syringe filter. Store at room temperature.

5. Thermal cycler (Mastercycler gradient, Eppendorf, Hamburg, Germany).

2.5 Aptamer Live Staining

1. 10 mg/ml sheared salmon sperm DNA. Store at −20 °C.

2. Blocking solution: Complete DMEM supplemented, 100 μg/ml sheared salmon sperm DNA described in the previous step (*see* **Note 7**). Prepare fresh before use.

3. Staining solution: Complete DMEM supplemented, 100 μg/ml sheared salmon sperm DNA containing, 100 nM folded aptamer. Prepare fresh before use.

4. Metal plate (e.g., length × width × thickness (cm): 20 × 12 × 2).

5. 4 % paraformaldehyde in 1× PBS solution (PFA): To prepare 1 L, add approximately 600 ml of 1× PBS to 40 g of PFA in a glass beaker and stir using a magnetic stirrer at ~50 °C. Adjust the pH between 7 and 8 using pH strips (do not contaminate the pH meter). Adjust the volume to 1 L with 1× PBS, make aliquots (e.g., 6 or 12 ml) and store at −20 °C. PFA is highly toxic: use gloves, lab coat, and respiratory and eye protection while handling PFA. Mix inside the fume hood. Please follow the waste bio safety recommendation of your institution before disposal of PFA.

6. DPBS: Store at room temperature.

7. Quenching solution: 0.1 M glycine in 1× DPBS. For 100 ml dilute 10 ml 1 M glycine stock solution (to prepare 100 ml of glycine 1 M, dissolve 7.5 g of glycine in ultrapure water; filter it through a 0.2 μm filter and store at room temperature) in 90 ml DPBS. Use freshly prepared working solution.

8. Mowiöl: For 100 ml of Mowiöl: mix with a magnetic stirrer 24 g glycerol, 9.6 g Mowiöl 4-88 reagent, 62.4 ml distilled

water and 9.6 ml of 1 M Tris buffer for 5–7 days. If necessary, heat the mixture up to 50 °C to help Mowiöl to dissolve. Make small aliquots (e.g., in 1 ml) and store at –20 °C until use.

2.6 Antibody and Transferrin Live Staining

1. Mouse monoclonal anti-transferrin receptor antibody (ab1086, Abcam, Cambridge, UK) (*see* **Note 8**).

2. Atto647N-labeled secondary antibodies anti-mouse IgGs.

3. Atto647N-coupled transferrin (specially made from Synaptic Systems, Göttingen, Germany).

4. 4 % PFA (*see* Subheading 2.5).

5. Primary antibody blocking solution: Complete DMEM supplemented, 2 % bovine serum albumin fraction V (BSA). Prepare fresh before use.

6. Primary antibody staining solution: Complete DMEM supplemented, 10 μg/ml mouse monoclonal anti-transferrin receptor antibody. Prepare fresh before use.

7. Metal plate (e.g., length × width × thickness (cm): 20 × 12 × 2).

8. DPBS: Store at room temperature.

9. Quenching solution: 0.1 M glycine in 1× DPBS. Store at room temperature until use.

10. Permeabilization and blocking solution (P&B solution): Prepare fresh before use 1× DPBS, 2 % BSA, and 0.1 % Triton X-100 (prepare a stock solution of 10 % Triton X-100). In a graduated cylinder pour 10 ml of Triton X-100 and fill it to 100 ml with water. Stir with a magnetic stirrer until solution is homogenous (it takes 2–5 h). Store at room temperature.

11. Secondary antibody incubation solution: Dilute the P&B solution 1:1 with DPBS. The antibody dilution needs to be empirically tested (here we use 1:100). Prepare fresh before use.

12. High-salt 1× DPBS (final concentration of 500 nM NaCl): Prepare 100 ml and store at room temperature.

13. Mowiöl (*see* Subheading 2.5).

2.7 Imaging

1. Microscope: Leica pulsed STED setup composed by a True Confocal System (TCS) STED SP5 fluorescence microscope equipped with a 100 × 1.4 NA HCX PL APO oil objective (Leica Microsystems GmbH, Mannheim, Germany) (*see* **Note 9**).

3 Methods

3.1 Dye Coupling to Aptamers (See Note 2)

1. Reduce 10 nmol of custom-synthesized aptamer bearing a C6-thiol group at the 5′ using 10 mM TCEP dissolved in 100 μl of TEAA.

2. Heat the sample for 3 min at 70 °C followed by 60 min at room temperature.

3. Desalt the reduced aptamers quickly into DPBS using Biospin 6 columns.

4. Dissolve the maleimide-functionalized fluorophore Atto647N in anhydrous DMSO to a concentration of 10 μg/μl (*see* **Note 10**).

5. Mix the reduced and desalted aptamer (~10 nmol) with 5 molar excess of functionalized dye (i.e., ~5 μl of the stock solution).

6. Incubate the coupling reaction overnight at 4 °C in dark conditions.

7. Recover the aptamers by ethanol precipitation, resuspend it in DPBS, and use a Biospin 6 column to desalt it.

8. To control the efficiency of the coupling reaction, run an analytical sample into a reversed phase HPLC. No unlabeled aptamer should be detected, and dye aptamer ratio under this coupling conditions should be typically ~1 (determined by absorbance at 260 and 650) (*see* **Note 11**).

3.2 Preparation of Glass Cover Slips

1. Place 18 mm glass cover slips in a beaker and cover them with 1 M HCl solution. Leave them shaking mildly (e.g., at 100 rpm) overnight at room temperature.

2. Wash cover slips 2 times with ultrapure water, cover them with 1 M NaOH, and leave them shaking mildly for 1 h at room temperature.

3. Wash with ultrapure water until water has a stable pH ~7.0. Place the cover slips in a new beaker covered with ethanol. Seal the beaker with Parafilm M and store at room temperature until use.

4. Under a sterile hood, use clean tweezers to take the cover slips, flame them one by one briefly and place them into the wells of a 12-well plate pre-filled with 1 ml of sterile water.

5. Aspirate the water and add 1 ml of 0.1 mg/ml poly-L-lysine (PLL) in every well containing a cover slip. Incubate for at least 2 h at room temperature.

6. Aspirate the PLL solution (*see* **Note 12**) and wash each well twice with 1 ml ultrapure sterile water.

7. Air-dry the plate inside the cell culture hood. Remove the lead of the plates and sterilize the open plates with ultraviolet light for 30–60 min. Store the plates and cover slips at 4 °C until use.

3.3 Cell Culture

1. A431 human cell line is grown in cell culture plates with complete DMEM medium incubated at 37 °C and 5 % CO_2.

2. Split and seed cells into 12-well plates containing the PLL-treated cover slips 1 day before the aptamer staining.

3. To split cells, first wash them thoroughly with sterile 1× DPBS (*see* **Note 13**). Add 2–3 ml of trypsin-EDTA to cover the surface of the plate and incubate for 1–5 min at 37 °C or until cells are completely detached. Stop trypsin activity by adding 10 ml of complete DMEM medium. Pipette up and down to get all cells from the plate and transfer them into a sterile 15 ml Falcon tube.

4. Centrifuge the cells with a swinging bucket rotor at 1000 rpm for 5 min at room temperature.

5. Aspirate the supernatant and resuspend the cell pellet in 10 ml of complete DMEM medium. Take 400 µl of the resuspended cells and add it to 12 ml of fresh complete DMEM medium. Add 1 ml of the later dilution into every well of the 12-well plates (*see* **Note 14**). The whole procedure must be performed in sterile conditions in a cell culture hood.

3.4 Preparation of Functional Aptamer

1. To prepare 10 µl of 10 µM properly folded aptamer (pre-working aptamer solution), mix 3.3 µl of fluorescently labeled aptamer (30 µM aptamer stock solution), 2 µl of 5× PBS, 2 µl of 5× MgCl$_2$, and 2.7 µl of DNase- and RNase-free distilled water (*see* **Note 6**).

2. Use a thermal cycler to heat up the pre-working aptamer solution to 75 °C for 3 min and then cool it down to 20 °C at a rate of 1 °C/min.

3.5 Aptamer Live Staining (See Note 15)

1. Aspirate the medium from the cells prepared during the step in Subheading 3.3, rinse briefly once with complete DMEM, and incubate for 10 min at 37 °C and 5 % CO$_2$ (cell culture incubator) with 300 µl of blocking solution per well.

2. In the meanwhile, fix with a tape a piece of Parafilm M on a metal plate. Make sure that the size of the Parafilm M is enough to fit all cover slips that are going to be stained. Preheat the metal plate and Parafilm M in the cell culture incubator.

3. Subsequently, carefully remove the cover slips from the 12-well plate (in blocking solution) using tweezers and place the cover slip upside down on 60 µl of staining solution spotted on the Parafilm M fixed to the preheated metal plate. Incubate the cells for 60 min inside the cell culture incubator (*see* **Note 16**).

4. For live imaging (Movie 1) carefully take the cover slip from the metal plate using tweezers. Remove the excess of staining solution by tapping the cover slip contour with a paper tissue and then submerge it several times into large volumes (e.g., 2 beakers containing ~100 ml each) of 1× DPBS at room temperature. Mount the cover slip in a live imaging chamber that fits the microscope stage and cover the cells with 400–500 µl of imaging DMEM medium.

5. For fixed samples (*see* Fig. 2): carefully take the cover slips from the metal plate using tweezers. Remove the excess of staining

solution by tapping the contour of the cover slip with a paper tissue, and then submerge it several times into large volumes (e.g., two beakers containing ~100 ml each) of ice-cold 1× DPBS (*see* **Note 17**). Briefly dry the cover slip by tapping it again on a tissue and place it into a well of a 12-well plate filled with 1 ml ice-cold 4 % PFA.

6. Keep the plate while fixing on ice for approximately 20 min and subsequently move it at room temperature for another 25 min (always keeping the sample under low light to avoid fluorophore bleaching).

7. Aspirate the PFA solution and add 1 ml of quenching solution to each well. Incubate for 15 min at room temperature in low light conditions.

8. Wash twice for 5 min with 1× DPBS (1 ml/well) and finally mount the cover slips with Mowiöl (e.g., 8 μl for 18 mm cover slips) in microscopy glass slides (*see* **Note 18**).

9. Dry the mounted cover slips in an oven at 37 °C for 15–20 min and store at 4 °C in the dark until imaged (*see* **Note 19**).

3.6 Antibody and Transferrin Live Stainings

1. Aspirate the medium from the cells, wash once with complete DMEM followed by 300 μl primary antibody blocking solution per well and incubate for 10 min at 37 °C and 5 % CO_2 (cell culture incubator).

2. In the meanwhile, fix with a tape a piece of Parafilm M on a metal plate. Make sure that the size of the Parafilm M is enough to fit all cover slips that are going to be stained. Preheat the metal plate and Parafilm M in the cell culture incubator.

3. Carefully remove the cover slips from the 12-well plate using tweezers and place the cover slip upside down on 60 μl of primary antibody staining solution spotted on the Parafilm M fixed to the preheated metal plate. Incubate the cells for 60 min in the cell culture incubator.

4. Take the cover slips carefully from the metal plate using tweezers. Remove the excess of staining solution by tapping the contour of the cover slip with a paper tissue and then submerge it several times into large volumes (e.g., two beakers containing ~100 ml each) of ice-cold 1× DPBS (*see* **Note 17**). Briefly dry the cover slip by tapping it again on a tissue and place it into a well of a 12-well plate filled with 1 ml ice-cold 4 % PFA.

5. Keep the plate while fixing on ice for approximately 20 min and subsequently move it at room temperature for another 25 min.

6. Aspirate the PFA solution and add 1 ml to each well of quenching solution (*see* Subheading 2.6). Incubate for 15 min at room temperature in the dark.

7. Wash once for 5 min with 1× PBS (1 ml/well).

8. Add 500 µl P&B solution per well and incubate for 20 min at room temperature.

9. Meanwhile prepare a staining chamber by fixing with a tape a piece of Parafilm M into a cell culture plate. Wrap the plate (lid and bottom separate) with aluminum foil and add inside wet paper tissues to the sides of the plate to keep a humid environment.

10. Remove the cover slips carefully from the 12-well plate using tweezers, and place the cover slip upside down on 60 µl of secondary antibody staining solution spotted on the Parafilm M fixed to the cell culture plate (prepared in previous step). Incubate the cells for 60 min at room temperature.

11. After incubation, remove the excess of staining solution by tapping the contour of the cover slip with a paper tissue and transfer the cover slips to a new 12-well plate containing 1 ml DPBS. Incubate at room temperature for 5 min protected from light.

12. Remove DPBS and add 1 ml per well of high-salt 1× DPBS (500 mM NaCl). Incubate for 10 min in dark conditions.

13. Wash three times for 5 min with 1× DPBS (1 ml/well).

14. Mount the cover slips in Mowiöl (e.g., 8 µl for 18 mm cover slips) (*see* **Note 18**).

15. Dry the mounted cover slips in an oven at 37 °C for 15–20 min and store at 4 °C in the dark until imaged.

3.7 Imaging

1. Images were acquired using a 100× 1.4 NA HCX PL APO oil objective. Excitation of Atto647N fluorophore was performed with a 635 nm pulsed laser (PicoQuant, Germany), and the 750 nm depletion STED beam was obtained with pulsed infrared titanium:sapphire tunable laser (Mai Tai Broadband, Spectra-Physics, Santa Clara, CA, USA).

2. For fixed samples (*see* Fig. 2) the pixel size was set to 20.2 nm, scanning speed was set to 1 kHz, lines were averaged 96 times, pinhole was set to one Airy unit, and signal was detected with an avalanche photodiode detector (APD).

3. STED and confocal images were acquired using the same parameters except that the STED depletion beam was off during confocal acquisitions.

4. For time-lapse STED imaging (Movie 1) pixel size was set to 20.2 nm, scanning speed was set to 8 kHz using a resonance scanner, pinhole was set to one Airy unit, and signal was detected with an avalanche photodiode detector (APD). Three images of 512 × 512 pixels were acquired per second for a period of 90 s.

4 Notes

1. Prepare all solutions using ultrapure water and analytical grade reagents (unless indicated otherwise). Prepare and store all reagents at room temperature and avoid light during steps with fluorophore-labeled aptamers (e.g., staining step and fixation) by covering tubes and plates with aluminum foil.

2. The c2 aptamer was custom synthesized in an automated synthesizer [17]. However, aptamers could be purchased from any company producing high-quality oligonucleotides (RNA and DNA). Fluorophore coupling is not particularly difficult, but it requires a high-performance liquid chromatography (HPLC) device. This step can be bypassed if aptamers are purchased already bearing a fluorescent moiety, thus completely avoiding the dye-coupling step described in Subheading 3.1.

3. Before using FCS to prepare the cell culture medium, it needs to be heat-inactivated. Incubate 500 ml of FCS in a water bath at 56 °C for 20 min, make aliquots in a sterile hood, and store them at –20 °C until use.

4. Alternatively, use any appropriate cell culture plates or flasks.

5. It is recommended to use the commercial sterile DPBS throughout the protocol.

6. If DNase- and RNase-free water is not purchased, it could be prepared as follows: add 50 μl of diethyl pyrocarbonate (DEPC) to 100 ml of ultrapure water (0.05 % final concentration). Mix and keep at room temperature overnight. Autoclave (to remove DEPC) and store at room temperature. DEPC is carcinogenic; therefore use proper personal protective equipment.

7. If unspecific binding of the aptamer persists, increase the concentration of sheared salmon sperm DNA and/or include equivalent amounts of yeast tRNA and/or use ~5 kDa dextran sulfate (in concentrations ranging from 0.1 to 1 mM).

8. For live stainings it is important that the antibody recognizes an extracellular domain of the receptor.

9. Here we describe the imaging of aptamers using laser scanning confocal and STED super-resolution microscopy. However, the staining procedure here presented should be suitable (with aptamers coupled to the appropriate fluorophores) for any other super-resolution microscopy technique. We have applied this staining methodology and successfully imaged using single-molecule localization microscopy [15, 19].

10. Maleimide group reacts with water molecules. Therefore, it is important to dissolve the dye in anhydrous DMSO and avoid water condensation or extended freeze-thaw cycles after making aliquots.

11. The absorbance at wavelength 260 nm monitors the concentration of oligonucleotide, and the second wavelength monitors the fluorophore. In our case, Atto647N has an efficient absorbance at ~650 nm; however, this needs to be adjusted depending on the particular fluorophore used.

12. PLL solution can be stored and used several times (4–6 times not older than 1 month). Store at 4 °C and always check the turbidity of the solution for obvious contamination before use.

13. Trypsin is rapidly inactivated in the presence of small traces of serum. Therefore, cell culture medium should be removed completely before treatment with trypsin by washing 1–3 times with DPBS if necessary.

14. Cells should cover 50–70 % (i.e., not crowded) of the cover slip surface before the staining (the day after splitting them). Therefore, the dilutions here proposed were optimized to the A431 cell line. However, if a different cell line is used, the dilution factors need to be determined according to their division rate to obtain this 50–70 % confluence at the time of staining.

15. The specificity of the c2 aptamer used here was thoroughly tested previously [15, 17]. However, it is extremely important to perform a negative control experiment to ensure that the aptamer is staining only the target of interest. Typically, this is achieved by performing a staining with a randomized sequence consisting of the same length and bearing the same fluorophore as the specific aptamer. Alternatively, staining cells lacking the target molecule can be used as negative control to ensure the aptamer specificity. However, this kind of control is difficult to perform if the target is a highly ubiquitous molecule (e.g., the transferrin receptor).

16. One-hour incubation with the aptamer was performed to achieve a full labeling of the trafficking pathway (due to the continuous cycling of the TfnRs). However, the incubation time with the aptamer should be adjusted depending of the target molecule, its biological properties (e.g., endocytosis mechanisms) and the question to study.

17. Washing in a large volume is performed to rapidly dilute the excess of aptamers in the sample. The use of ice cold PBS is necessary to stop or slow down the movement of vesicles and endosomes and "freeze" the trafficking system in the cells.

18. Place the cover slip carefully in a Mowiöl drop, avoiding the formation of bubbles. Do not press the cover slip once it is in Mowiöl.

19. Aptamer stained samples embedded in Mowiöl can be stored at 4 °C for 1 week. However, the best results are obtained when samples are imaged directly or the next day after mounting them.

Acknowledgments

We thank Sven Truckenbrott, Sinem K. Saka, Silvio Rizzoli, and Corinna Opazo for carefully reading and adjusting the manuscript.

References

1. Abbe E (1873) Beiträge zur Theorie des Mikroskops und der mikroskopischen Wahrnehmung. Arch Mikrosk Anat 9(1):413–468

2. Hell SW, Wichmann J (1994) Breaking the diffraction resolution limit by stimulated emission: stimulated-emission-depletion fluorescence microscopy. Opt Lett 19:780–782

3. Betzig E, Patterson GH, Sougrat R et al (2006) Imaging intracellular fluorescent proteins at nanometer resolution. Science 313:1642–1645. doi:10.1126/science.1127344

4. Rust MJ, Bates M, Zhuang X (2006) Imaging by stochastic optical reconstruction microscopy (STORM). Nat Methods 3:793–795. doi:10.1038/NMETH929

5. Gustafsson MG (2000) Surpassing the lateral resolution limit by a factor of two using structured illumination microscopy. J Microsc 198:82–87

6. Fölling J, Bossi M, Bock H et al (2008) Fluorescence nanoscopy by ground-state depletion and single-molecule return. Nat Methods 5:943–945. doi:10.1038/NMETH.1257

7. Schermelleh L, Heintzmann R, Leonhardt H (2010) A guide to super-resolution fluorescence microscopy. J Cell Biol 190:165–175. doi:10.1083/jcb.201002018

8. Fornasiero EF, Opazo F (2015) Super-resolution imaging for cell biologists: concepts, applications, current challenges and developments. Bioessays 37:436–451. doi:10.1002/bies.201400170

9. Willig KI, Rizzoli SO, Westphal V et al (2006) STED microscopy reveals that synaptotagmin remains clustered after synaptic vesicle exocytosis. Nature 440:935–939. doi:10.1038/nature04592

10. Zhang WI, Röhse H, Rizzoli SO, Opazo F (2014) Fluorescent in situ hybridization of synaptic proteins imaged with super-resolution STED microscopy. Microsc Res Tech 77:517–527. doi:10.1002/jemt.22367

11. Xu K, Zhong G, Zhuang X (2013) Actin, spectrin, and associated proteins form a periodic cytoskeletal structure in axons. Science 339:452–456. doi:10.1126/science.1232251

12. Opazo F, Punge A, Bückers J et al (2010) Limited intermixing of synaptic vesicle components upon vesicle recycling. Traffic 11:800–812. doi:10.1111/j.1600-0854.2010.01058.x

13. Fernández-Suárez M, Ting AY (2008) Fluorescent probes for super-resolution imaging in living cells. Nat Rev Mol Cell Biol 9:929–943. doi:10.1038/nrm2531

14. Ries J, Kaplan C, Platonova E et al (2012) A simple, versatile method for GFP-based super-resolution microscopy via nanobodies. Nat Methods 9:582–584. doi:10.1038/nmeth.1991

15. Opazo F, Levy M, Byrom M et al (2012) Aptamers as potential tools for super-resolution microscopy. Nat Methods 9:938–939. doi:10.1038/nmeth.2179

16. Endesfelder U, Heilemann M (2014) Art and artifacts in single-molecule localization microscopy: beyond attractive images. Nat Methods 11:235–238. doi:10.1038/nmeth.2852

17. Wilner SE, Wengerter B, Maier K et al (2012) An RNA alternative to human transferrin: a new tool for targeting human cells. Mol Ther Nucleic Acids 1, e21. doi:10.1038/mtna.2012.14

18. Richardson DR, Ponka P (1997) The molecular mechanisms of the metabolism and transport of iron in normal and neoplastic cells. Biochim Biophys Acta 1331:1–40

19. Egner A, Geisler C, von Middendorff C et al (2007) Fluorescence nanoscopy in whole cells by asynchronous localization of photoswitching emitters. Biophys J 93:3285–3290. doi:10.1529/biophysj.107.112201

20. Wilhelm BG, Mandad S, Truckenbrodt S et al (2014) Composition of synaptic boutons reveals the amounts of vesicle trafficking proteins. Science 344:1023–1028

Chapter 18

Synthesis and Characterization of Aptamer-Targeted SNALPs for the Delivery of siRNA

Samantha E. Wilner and Matthew Levy

Abstract

Aptamers selected against cell surface receptors represent a unique set of ligands that can be used to target nanoparticles and other therapeutics to specific cell types. Here, we describe a method for using aptamers to deliver stable nucleic acid lipid particles (SNALPs) encapsulating small interfering RNA (siRNA) to cells in vitro. Using this method, we have demonstrated the ability of aptamer-conjugated SNALPs to achieve target-specific delivery and siRNA-mediated knockdown of a gene of interest. We also describe methods to characterize SNALP size, siRNA encapsulation efficiency, and aptamer conjugation efficiency.

Key words Aptamer, SNALP, siRNA, Liposome, Targeted delivery

1 Introduction

Small interfering RNAs (siRNAs) have tremendous therapeutic potential because they can be designed to target any gene of interest, resulting in mRNA degradation and reduced expression of their target protein. Practical use of siRNA in the clinic, however, has met many challenges such as poor delivery of siRNA to target tissues, difficulties determining appropriate dosage, and rapid siRNA degradation by nucleases in vivo. In addition to chemical modification of the siRNA itself, delivery methods involving nanoparticles and liposomes have been developed to shield siRNA from nucleases and allow for siRNA to cross the cell membrane in vivo.

One promising platform for the delivery of siRNA is a subset of liposomes called stable nucleic acid lipid particles (SNALPs). SNALPs are characterized by high siRNA encapsulation efficiency, allowing for lower particle dosing to achieve desired gene knockdown. SNALP composition further enhances the delivery properties of these particles because they consist of a novel combination of cationic and fusogenic lipids [1, 2]. These lipids promote fusion of the liposomal membrane with either the endosome or cell membrane, allowing for release of siRNA in the

Günter Mayer (ed.), *Nucleic Acid Aptamers: Selection, Characterization, and Application*, Methods in Molecular Biology, vol. 1380, DOI 10.1007/978-1-4939-3197-2_18, © Springer Science+Business Media New York 2016

cytoplasm. Furthermore, SNALPs contain a polyethylene glycol (PEG)-modified lipid that reduces SNALP aggregation and increases particle circulation times in vivo by forming a hydrophilic layer around the liposome [3].

Passive targeting of SNALPs has been shown to result in particle accumulation in the liver. Active targeting ligands, however, can be introduced to the particle surface to enhance cell uptake in tissues of interest [4]. Numerous different targeting ligands are available to alter the targeting specificity of SNALPs including, antibodies, antibody fragments (e.g., single-chain antibodies, diabodies) peptides and small molecules [5–10]. Cell-type-specific aptamers are another such tool that can be used for this purpose [11–15]. Here, we describe a method used for preparing and characterizing SNALPs that are modified with aptamers, which target cell surface receptors.

As an example we describe the use of an aptamer (c2.min) which targets the human transferrin receptor (TfR) to enhance siRNA uptake leading to increased target gene knockdown [11]. Along with characterizing SNALP size, aptamer conjugation, and siRNA encapsulation efficiency, we also describe experiments to assess SNALP uptake and the extent to which SNALPs knock down their respective gene targets. For this we make use of two distinct assay formats. Using HeLa cells constitutively expressing EGFP (HeLa-EGFPs) we assess both cell uptake and protein knockdown by flow cytometry.

Finally, we note that while the protocol detailed here is specific for synthesizing aptamer targeted SNALPs, a similar methodology can readily be applied to other liposome formulations as well as other nanoparticles.

2 Materials

2.1 siRNA Preparation in Citrate Buffer

1. Anti-EGFP siRNA duplex made of antisense sequence (5′ UGCGCUCCUGGACGUAGCCTT 3′) and sense sequence (5′ GGCUACGUCCAGGAGCGCATT 3′) where uridine and cytidine are 2′-fluoro modified. Both 3′ overhangs (TT) are made of deoxythymidine.

2. Phosphate-buffered saline with Mg^{2+} and Ca^{2+} (DPBS(+), 10×, pH 7.4): Weigh 1 g calcium chloride ($CaCl_2$) anhydrous, 1 g magnesium chloride ($MgCl_2 \cdot 6H_2O$), 2 g potassium chloride (KCl), 2 g potassium phosphate monobasic (KH_2PO_4), 80 g sodium chloride (NaCl), 21.6 g sodium phosphate dibasic ($Na_2HPO_4 \cdot 7H_2O$). Make up to 1 L with water and filter through a 0.45 µm filter.

3. 200 mM Citrate buffer, pH 5.0): Weigh 1.92 g citric acid anhydrous. Make up to 50 mL with water and 1.95 mL sodium hydroxide.

4. Micro Bio-Spin 6 columns.

5. 12×75 mm tubes.

6. Nanodrop

2.2 Thiol-Aptamer Reduction and Conjugation

1. 2 M Triethylammonium acetate (TEAA, 20×): 57.5 mL acetic acid and 139.3 mL triethylammonium acetate in 303.2 mL water.

2. 500 mM Tris-(2-carboxyethyl) phosphine (TCEP).

3. Micro Bio-Spin 6 columns

4. 12×75 mm round-bottom tubes

5. Phosphate-buffered saline (DPBS(-), 10×, pH 7.4): Weigh 2 g potassium chloride (KCl), 2 g potassium phosphate monobasic (KH$_2$PO$_4$), 11.4 g sodium phosphate dibasic anhydrous (Na$_2$HPO$_4$), and 80 g sodium chloride (NaCl). Make up to 1 L with water and filter through a 0.45 μm filter.

6. Nanodrop

7. ß-Mercaptoethanol

2.3 Preparation of Lipid Mixture

1. DSPC (1,2-distearoyl-sn-glycero-3-phosphocoline).

2. DSPE-PEG-Mal (1,2-distearoyl-sn-glycero-3-phosphoethanolamine-*N*-[maleimide(polyethylene glycol)-2000] (Avanti Polar Lipids, Alabaster, AL).

3. DLinDMA (1,2-dilinoleyloxy-*N*,*N*-dimethyl-3-aminopropane, *see* **Note 1**).

4. Cholesterol.

5. Ethanol (200 proof).

6. Chloroform.

7. 10×75 mm glass tubes.

8. Speed vacuum.

2.4 SNALP Preparation

1. siRNA (*see* Subheading 2.1).

2. Lipid mixture (*see* Subheading 2.3).

3. Heat block.

4. Ethanol (200 proof)

5. Sterile water

6. 200 mM Citrate buffer (10×, pH 5.0): Weigh 1.92 g citric acid anhydrous. Make up to 50 mL with water and 1.95 mL 10 N sodium hydroxide.

7. 200 mM Citrate buffer (10×, pH 6.0): Weigh 1.92 g citric acid anhydrous. Make up to 50 mL with water and 2.655 mL 10 N sodium hydroxide.

8. Citrate buffer with NaCl: 1× citrate buffer, pH 6.0, 300 mM sodium chloride.

9. Slide-A-Lyzer MINI Dialysis Units, 10 kDa

10. Phosphate-buffered saline (DPBS(–), 25×, pH 8.0): Weigh 33.8 g sodium phosphate dibasic anhydrous, 1.9 g sodium phosphate monobasic dihydrate, and 218 g sodium chloride. Make up to 1 L with water and filter through a 0.45 μm filter.

11. Dialysis buffer: 1× DPBS(–) (pH 8.0) supplemented with 1 mM EDTA

2.5 Assaying SNALP Size

1. Dynamic light scattering instrument such as DynaPro Plate Reader (Wyatt Technology, Santa Barbara, CA, USA).

2. Phosphate-buffered saline (DPBS(–), 1×, pH 7.4).

2.6 Assaying siRNA Encapsulation

1. Quan-iT RiboGreen (Invitrogen, Carlsbad, CA, USA).

2. TritonX-100.

3. TE buffer: 10 mM Tris–HCl, pH 7.5, 1 mM EDTA.

4. TE/TritonX buffer: 10 mM Tris–HCl, pH 7.5, 1 mM EDTA, 0.5 % TritonX-100.

5. RiboGreen assay buffer: TE buffer with RiboGreen (1:500 dilution).

6. RiboGreen assay buffer with TritonX: TE/TritonX buffer with RiboGreen (1:500 dilution).

7. White opaque 96-well plate.

8. Fluorescent plate reader such as the Synergy H4 Hybride Multi-Mode Microplate Reader (BioTek, Winooski, VT, USA).

2.7 Assaying Aptamer Conjugation

1. Electrophoresis system.

2. Tris-borate-EDTA buffer (TBE, 10×): 890 mM Trizma base, 890 mM boric acid, 20 mM EDTA, pH 8.0.

3. 12 % Denaturing polyacrylamide solution, filtered: 300 mL 40 % 19:1 acrylamide:bis-acrylamide, 7 M urea, 100 mL 10× TBE, dH$_2$O up to 1 L.

4. 2× Stop dye: 95 % formamide, 20 mM EDTA, pH 8.0, 0.1 % bromophenol blue.

5. 10 % Ammonium persulfate (APS).

6. Tetramethylethylenediamine (TEMED).

7. O'RangeRuler 20 bp DNA ladder.

8. SYBR Gold Nucleic Acid Gel Stain.

9. Fluorescent imaging system such as the STORM 840 phosphorimager/fluorimager (GE Healthcare, Pittsburgh, PA, USA).

2.8 Assaying SNALP Uptake by Flow Cytometry

1. HeLa cells constitutively expressing enhanced green fluorescent protein (EGFP, *see* **Note 2**).

2. D5: DMEM media supplemented with 5 % heat inactivated fetal bovine serum (FBS).

3. D5 blocked media: DMEM media supplemented with 5 % FBS, 1 mg/mL tRNA, and 1 mg/mL ssDNA.

4. Selective antibiotic such as G418.

5. 0.05 % Trypsin in 0.53 mM EDTA.

6. DPBS without Mg^{2+} or Ca^{2+} (Invitrogen, Carlsbad, CA, USA).

7. 5 % CO_2 incubator at 37 °C and 99 % humidity.

8. 96-Well tissue culture treated plates.

9. 25 cm^2 tissue culture treated flasks.

10. FACS buffer: 1 % bovine serum albumin (BSA), 0.1 % sodium azide in HBSS.

11. FACS buffer with 5 μg/mL bisbenzimide.

3 Methods

3.1 siRNA Preparation in Citrate Buffer

1. If not already double stranded, anneal sense strand to its complementary antisense sequence. Add sense and antisense strands at equimolar quantities in 50–100 μL of 1× DPBS(+). Incubate at 70 °C for 3 min. Let cool at room temperature for 15 min. The formation of the siRNA duplex can be confirmed by analysis on an agarose gel (*see* **Note 3**).

2. Prepare the Micro Bio-Spin 6 following the manufacturer's protocol. Invert column to resuspend the packed gel and remove bubbles. Snap off the end of the column and remove the column cap. Place the column in a 12×75 mm tube. Centrifuge at 1000×*g* for 1 min in a spin-bucket centrifuge. Discard flow through.

3. Wash the column with 20 mM sodium citrate buffer (pH 5.0). Add 1 mL of sodium citrate buffer to the column. Centrifuge at 1000×*g* for 1 min. Discard the flow through and repeat this step. Add another 1 mL of citrate buffer (pH 5.0) and centrifuge at 1000×*g* for 2 min.

4. Transfer column to a 1.5 mL Eppendorf. Apply siRNA (50–100 μL) to the top of the column. Centrifuge at 1000×*g* for 4 min.

5. Quantify siRNA using a Nanodrop or another UV spectrometer. Use a 1:10 dilution of the recovered RNA in DPBS(−) (*see* **Note 4**).

3.2 Thiol-Aptamer Reduction

1. Resuspend thiolated aptamer in 0.1 M TEAA.

2. Add TCEP to a final concentration of 10 mM (*see* **Note 5**).

3. Heat aptamer at 70 °C for 3 min.

4. Incubate sample at room temperature for a minimum of 1 additional hour.

5. Remove TCEP using a micro Bio-Spin 6 column. Prepare the column as instructed in Subheading 3.1, **step 2**.

6. Wash the column with DBPS(–). Add 1 mL of DPBS(–) to the column. Centrifuge at $1000 \times g$ for 1 min. Discard the flow through and repeat this step. Add another 1 mL of DPBS(–) and centrifuge at $1000 \times g$ for 2 min.

7. Transfer column to a 1.5 mL Eppendorf. Apply aptamer (50–100 μL) to the top of the column. Centrifuge at $1000 \times g$ for 4 min.

8. Quantify aptamer using a Nanodrop or another UV spectrometer.

9. Store at –80 °C until needed (*see* **Note 6**, regarding the storage of thiolated reduced aptamers).

3.3   Preparation of Lipid Mixture

1. Prepare lipid stocks by resuspending DLinDMA, cholesterol, and DSPE-PEG-Mal, respectively, at 10 mg/mL and DSPC at 50 mg/mL in chloroform (*see* **Note 1**, regarding the synthesis of DLinDMA). Store all lipid chloroform stocks in glass vials topped with argon at –80 °C.

2. Mix lipids together at a desired molar ratio in a glass tube. For effective in vitro knock down, a molar ratio of 10:40:48:2 DSPC:DLinDMA:cholesterol:DSPE-PEG-Mal is recommended (*see* **Note 7**). Prepare a 10 μmol stock of lipid mixture by mixing 15.8 μL DSPC (at 50 mg/mL), 246.4 μL DLinDMA (at 10 mg/mL), 185.6 μL cholesterol (at 10 mg/mL), and 58.8 μL DSPE-PEG-Mal (at 10 mg/mL).

3. Concentrate the lipid mixture in a speed vacuum until the chloroform has been completely removed.

4. Resuspend the dried lipid mixture at 10 mg/mL in 100 % ethanol. For the recommended lipid mixture in **step 2**, add 570 μL ethanol.

5. Aliquot the lipid mixture into 25 μL volumes and lyophilize for storage purposes. Store lipid aliquots at –80 °C.

3.4   SNALP Preparation

1. Resuspend a lipid mixture aliquot prepared in Subheading 3.3 by adding 22.5 μL ethanol plus 2.5 μL of water to make a 25 μL 90 % vol/vol ethanol solution. Each lipid mix will be at 10 mg/mL and will contain 250 μg of lipid.

2. Prepare a second solution of 25 μL containing siRNA at a 1:10 M ratio of fluorescently labeled siRNA to unlabeled siRNA in 20 mM citrate buffer, pH 5.0. The amount of siRNA added should maintain a 3:1 (+/–) charge ratio based on the molar amounts of negatively charged siRNA and positively charge DLinDMA. Following the suggested protocol, ~1.4

nmol of siRNA needs to be added to achieve a 3:1 (+/−) ratio. Use 20 mM citrate buffer, pH 5.0, to bring the volume of the solution to 25 μL.

3. Heat both the lipid solution and siRNA solutions separately to 37 °C.

4. Add the siRNA solution to the lipid solution with rapid mixing (*see* **Note 8**).

5. Immediately dilute the lipid/siRNA mixture with 50 μL of 20 mM citrate buffer, pH 6.0 containing 300 mM NaCl, pre-heated to 37 °C. Mix well by rapid pipetting.

6. Incubate the final SNALP solution at 37 °C for 30 min.

7. Dialyze the SNALP mixture overnight at 4 °C with a 10 kDa membrane into DPBS(−), pH 8.0, with 1 mM EDTA.

3.5 Aptamer Conjugation

1. Following dialysis, aliquot liposome mixture into equal volumes for conjugation to aptamer. It is a good idea to always retain one aliquot that will not be conjugated to aptamer as a control.

2. Add reduced thiol-modified aptamers to the appropriate liposome samples. The number of aptamers per liposome can be varied by altering the ratio at this step. The number of liposomes per μL of solution can be determined by assuming unilamellar liposomes, an average lipid surface area of 0.6 nm², and a mean particle diameter of 180 nm, which can be measured using DLS [16]. In the example above, 250 μg of lipid in 100 μL, gives ~1.6×10^{10} particles/μL. For our experiments, we set the aptamer to liposome ratio at ~60 aptamers per liposome (assuming 100 % conjugation; *see* **Note 9**). However, this value can be increased significantly as there are ~3300 DSPE-PEG-Mal molecules per liposome.

3. Incubate for >3 h at room temperature or overnight at 4 °C.

4. Quench unreacted maleimide groups with 1 mM BME (final concentration). Incubate for ~2 h at room temperature.

5. Aptamer-functionalized and nonfunctionalized SNALPs can be stored at 4 °C for 7 days.

3.6 Assaying SNALP Size

Dynamic light scattering (DLS) can measure particle diameter and size distribution (*see* **Note 10**).

1. Prepare a dilution series of liposomes in DPBS(−) for measurement via DLS. Optimal liposome concentration will vary based on the DLS instrument. Using the DynaPro Plate Reader, we have determined that a lipid concentration of ~50 μg/mL is an appropriate concentration for measuring liposome diameter. By measuring particle diameter over a dilution series, it will be possible to assess the monodispersity of the population. Monodisperse particles will show little variation in mean

diameter over the concentration range, within the limits of detection of the instrument.

2. Follow the manufacturer's protocol to determine liposome diameter.

3.7 Assaying siRNA Encapsulation

siRNA encapsulation can be determined using RiboGreen [2, 17].

1. Prepare siRNA stocks in TE buffer to be used as a standard curve at the following concentrations: 100 nM, 50 nM, 25 nM, 12.5 nM, 6.25 nM, and 3.125 nM. Use the same siRNA that was used to prepare liposomes. For example, if using siRNA at a 1:10 M ratio of Cy5-labeled siRNA to unlabeled siRNA, make each siRNA stock at the identical molar ratio with the same original stocks of siRNA (both labeled and unlabeled). Failure to do so may give inaccurate results.

2. Prepare another set of siRNA stocks at the same concentrations as **step 1** in TE/TritonX buffer.

3. Make two standard curves by adding 100 μL of each siRNA stock (*see* **steps 1** and **2**) to a white opaque 96-well plate (Fig. 1a).

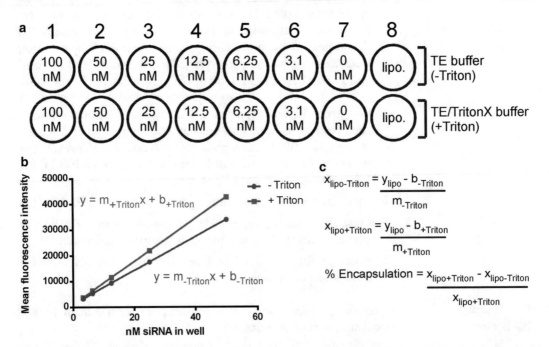

Fig. 1 Example of analysis for the RiboGreen assay to assess siRNA encapsulation. (**a**) Representation of a 96-well plate setup consisting of two siRNA standard curves, one prepared in TE buffer without TritonX (-Triton) and one prepared in TE buffer with TritonX (+Triton). The presence of TritonX changes the emission intensity of RiboGreen; thus we recommend preparing two standard curves in this manner. (**b**) Plot of mean fluorescence intensity versus siRNA concentration per well. Determine the best-fit line for each standard curve, where *m* is the slope, *b* is the *y*-intercept, *x* is siRNA concentration per well, and *y* is the mean fluorescence intensity. (**c**) Equations used to calculate the percent of siRNA encapsulation

4. Add 100 μL of TE buffer and 100 μL TE/TritonX buffer as "blank" samples to the 96-well plate.

5. Prepare the SNALP sample (with no aptamer) by first diluting the sample 1:2. Next, add 1 μL of the diluted liposome to both 100 μL of TE buffer and 100 μL of TE/TritonX buffer, respectively, in the white opaque 96-well plate. Do not use an aptamer-conjugated SNALP sample because the aptamer will bind RiboGreen and skew the results.

6. Add 100 μL of the appropriate RiboGreen assay buffer to each well. All wells with TE/TritonX buffer should get RiboGreen assay buffer containing TritonX.

7. Incubate the plate for 5 min at room temperature in the dark.

8. Excite samples at 485 nm and measure the fluorescence emission intensity at 530 nm using a fluorescence microplate reader.

9. To analyze the results, subtract the fluorescence emission intensity of the background sample from that of each of the samples. Plot the fluorescence emission intensity versus siRNA concentration to make a standard curve for the samples prepared in TE buffer and for those prepared in TE/TritonX buffer, respectively. Determine the equation of the best-fit line for these two curves, and use these equations to determine the siRNA concentration of the liposome sample both in TE buffer and in TE/TritonX buffer. The siRNA concentration of the liposome in TE/TritonX buffer represents total siRNA whereas that in TE buffer represents the non-encapsulated siRNA. Determine the percentage of encapsulated siRNA by subtracting the amount of non-encapsulated siRNA from the total and then dividing by the total siRNA concentration (Fig. 1b, c).

3.8 Assaying Aptamer Conjugation

The extent of aptamer conjugation can be determined by gel electrophoresis.

1. Assemble glass plates for polyacrylamide gel electrophoresis (*see* **Note 11**).

2. Make the polyacrylamide solution by mixing 7 mL 12 % denaturing polyacrylamide solution, 23.5 μL 10 % APS, and 5 μL TEMED.

3. Mix the polyacrylamide solution well and pipet slowly into the glass plates.

4. Insert at 10-well comb and allow the gel to polymerize completely before use.

5. While the gel is polymerizing, prepare the 20 bp ladder by diluting 1 μL of ladder in 4 μL of PBS.

6. Prepare samples for a standard curve, each in 5 μL total volume, by adding reduced aptamer at the following amounts: 1, 5, 10, 20, and 30 ng.

7. Prepare the liposome sample by diluting 1 μL of liposome sample into 4 μL of PBS.

8. Add 5 μl of 2× stop dye to all samples, bringing the sample volume to 10 μL.

9. Boil samples at 70 °C for 3 min.

10. Once the gel is polymerized, remove the comb from the gel and rinse briefly with water to remove excess acrylamide. Assemble the gel into an electrophoresis system.

11. Heat 1× TBE to ~70 °C and use this to fill the electrophoresis system. This can be done by warming 1× TBE in a microwave to ~70 °C (*see* **Note 11**).

12. Before adding samples to the gel, remove leached urea from each well using a pipette.

13. Load samples into the well.

14. Electrophorese at ~215 V until the bromophenol blue in the stop dye reaches the bottom of the gel.

15. Gently remove the gel from the glass plates. Place the gel in ~25 mL of distilled water in a shallow basin. Place the gel basin on a slow shaker for 5 min to rinse the gel.

16. Discard the water and replace with ~25 mL of 1× TBE with 1× SYBR Gold. Place the gel basin back on the shaker and incubate for 15 min to stain for RNA.

17. Wash off excess SYBR Gold by rinsing the gel three times with distilled water.

18. Image the gel using a fluorescence imaging system (Fig. 2).

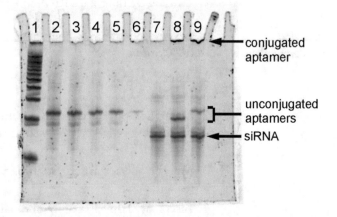

Fig. 2 Representative 12 % denaturing acrylamide gel to measure aptamer conjugation. Standard curve in *lanes 2–6* was made using anti-human transferrin receptor aptamer, c2.min. *Lane 1*: 20 bp ladder. *Lane 2*: 30 ng aptamer. *Lane 3*: 20 ng aptamer. *Lane 4*: 10 ng aptamer. *Lane 5*: 5 ng aptamer. *Lane 6*: 1 ng aptamer. *Lane 7*: SNALPs without aptamer. *Lane 8*: SNALPs conjugated to control aptamer. *Lane 9*: SNALPs conjugated to c2.min

19. Quantify the fluorescence intensity of each band. Subtract the background intensity of the gel from each sample. Plot aptamer concentration versus the background subtracted intensity to make a standard curve. Use the best-fit line for this curve to determine the percent aptamer conjugation (*see* **Note 9**).

3.9 Assaying SNALP Uptake by Flow Cytometry

1. Seed 25,000 HeLa-EGFP cells per well in a 24-well plate in 1 mL D5 media. Allow the cells to grow overnight at 37 °C with 5 % CO_2 and 99 % humidity.

2. Remove media and replace with 1 mL blocked D5 media to each well. Incubate for 1 h at 37 °C with 5 % CO_2 and 99 % humidity.

3. Add each SNALP sample at a final concentration of 200 nM siRNA per well, based on the amount of siRNA encapsulation determined using the RiboGreen assay.

4. Incubate for 24 h at 37 °C.

5. Replace media in all wells with fresh D5 media and further incubate for 24 h before assaying.

6. Remove media and wash cells with DPBS(−) by gentle pipetting.

7. Add 200 μL of trypsin to lift cells followed by incubation at 37 °C for 5–10 min.

8. Once cells have lifted from the plate, quench trypsinization by adding 1 mL of FACS buffer.

9. Transfer cells to a 1.5 mL microcentrifuge tube.

10. Centrifuge cells at $300 \times g$ for 3 min.

11. Remove supernatant and resuspend cells in 500 μL FACS buffer with bisbenzimide.

12. Transfer samples to FACS tubes.

13. Analyze samples by flow cytometry to assess siRNA uptake and GFP knockdown (Fig. 3).

4   Notes

1. DLinDMA is not commercially available but can be synthesized according to the protocol published by Heyes et al. [2] Other commercially available lipids such as DODAP (1,2-dioleoyl-3-dimethylammonium-propane) can also be used; however in our hands, this molecule fails to yield SNALPs capable of effectively delivering siRNA. More recently, Jayaraman et al. identified additional cationic lipids which improve siRNA efficacy in vivo. Although these lipids are not commercially

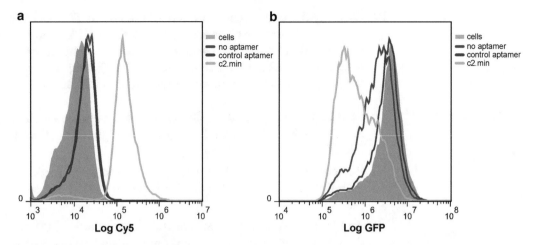

Fig. 3 Representative flow cytometry data to assess SNALP uptake and siRNA-mediated knockdown. (**a**) SNALP uptake by HeLa-EGFP cells. Cells were incubated with SNALPs as described in Subheading 3.9. SNALPs conjugated to c2.min show specific uptake (*green*) whereas SNALPs bearing a negative control aptamer (*blue*) or those with no aptamer (*purple*) show little uptake. (**b**) Knockdown of EGFP expression. Targeted SNALPs (*green*) show an increased amount of EGFP knockdown compared to negative controls (*blue* and *purple*)

available, they can be synthesized according to the protocol published by Jayaraman et al. [18].

2. For these experiments, we utilize a HeLa cell line which expresses a destabilized variant of EGFP which has a half-life of ~6 h.

3. The formation of the siRNA duplex can be confirmed by analysis on an agarose gel, running ~100 ng of the annealed material on a 4 % agarose gel followed by staining with ethidium bromide. Single-stranded RNAs should be analyzed for comparison as well as a ladder. For this purpose, a dsDNA ladder such as the O'RangeRuler 20 base pair DNA ladder (Thermo Scientific, Waltham, MA, USA) is sufficient to confirm the approximate size of the duplexed RNA which runs roughly equivalent to a dsDNA duplex.

4. Dilution of the siRNA in DPBS(–) is necessary to adjust the pH to ~7.4 and thus more accurately assess the concentration of the siRNA.

5. The literature states that TCEP can be stored for extended periods at room temperature; however, we have seen numerous aqueous stocks go bad. We now buy TCEP in 500 mM aqueous stocks in 1 mL ampules (Sigma-Aldrich, St. Louis, MO, USA). Upon opening an ampule, we transfer the solution to a 1 mL screw cap tube and store it at –20 °C between uses. Our standard thiol-reduction protocol using 0.1 M TEAA almost always gives complete reduction (provided the TCEP has not gone bad). Other buffers can be used, but we have found the

most consistent results in 0.1 M TEAA. Do not use phosphate buffers. Reduction can be checked by reverse-phase HPLC as the reduced aptamer has a shorter retention time than the unreduced molecule. For reverse-phase HPLC, we typically employ a 4.6×10 mm Xbridge BEH C18 column (Waters Corporation, Milford, MA, USA) heated to 65 °C and a gradient of acetonitrile (ACN) in 0.1 M TEAA, pH 7.5. Our standard gradient is 10 % ACN to 20 % ACN over 5 min followed by an increase to 70 % ACN over 10 min.

6. We typically reduce enough aptamer for a typical experiment prior to conjugation to ensure that it is reduced. However, we have found that reduced aptamer can be stored at −80 °C for extended periods of time (weeks to months) without the formation of disulfide-linked dimers. Care should be taken to avoid prolonged storage (>24 h) at 4 °C or storage in frost-free −20 °C freezers as this can allow for dimerization. Dimerization can be readily detected on an acrylamide gel, and the disulfide-linked molecules subsequently re-reduced and desalted.

7. The molar ratios of lipids may affect in vitro siRNA-mediated knockdown. Specifically in our hands, increasing DSPE-PEG-Mal above 2 mol% diminishes siRNA functionality. Although we are able to see targeted uptake using greater amounts of DSPE-PEG-Mal, we do not see knockdown.

8. Rapid mixing can be performed for small-scale reactions like those described here using simple pipetting. We have successfully scaled the method to volumes as large as ~100 μL. For larger volumes, it is advisable to utilize a T-mixer (such as micro-static mixing tee #U-466, Idex, Health Science, Oak Harbor, WA, USA) and two syringes. Readers should refer to Jeffs et al. for more elaborate flow and mixing systems [19].

9. Although aptamer conjugation to liposomes is not 100 % efficient, we do not clean up unconjugated aptamer from the final liposome preparation. For this reason, it is important to check aptamer conjugation efficiency by gel electrophoresis. Typically, aptamer conjugation is ~90 % efficient, which leaves little unreacted aptamer that could interfere with future experimental results.

10. It is a good idea to use DLS to determine liposome size both before and after aptamer conjugation to the liposomes to ensure that their addition is not dramatically altering size. Additionally, determining the size of the liposomes before the addition of the aptamers is necessary to estimate the appropriate amount of aptamer to utilize during the conjugation step.

11. For analytical denaturing (7M urea) gels, we typically use 1.0 mm Bio-Rad mini-PROTEAN gel plates (Bio-Rad, Hercules, CA, USA). Because the buffer reservoir runs the

length of the glass plate, we typically heat the buffer to 70 °C before filling the apparatus. This aids greatly in denaturing the samples and is essential for fully denaturing the 20 bp dsDNA ladder.

References

1. Semple SC, Akinc A, Chen J et al (2010) Rational design of cationic lipids for siRNA delivery. Nat Biotechnol 28:172–176

2. Heyes J, Palmer L, Bremner K et al (2005) Cationic lipid saturation influences intracellular delivery of encapsulated nucleic acids. J Control Release 107:276–287

3. Heyes J, Hall K, Tailor V et al (2006) Synthesis and characterization of novel poly(ethylene glycol)-lipid conjugates suitable for use in drug delivery. J Control Release 112:280–290

4. Lammers T, Kiessling F, Hennink WE et al (2011) Drug targeting to tumors: principles, pitfalls and (pre-) clinical progress. J Control Release

5. Davis ME, Zuckerman JE, Choi CHJ et al (2010) Evidence of RNAi in humans from systemically administered siRNA via targeted nanoparticles. Nature 464:1067–1070

6. Davis ME (2009) The first targeted delivery of siRNA in humans via a self-assembling, cyclodextrin polymer-based nanoparticle: from concept to clinic. Mol Pharm 6:659–668

7. Akinc A, Querbes W, De S et al (2010) Targeted delivery of RNAi therapeutics with endogenous and exogenous ligand-based mechanisms. Mol Ther 18:1357–1364

8. Sato Y, Murase K, Kato J et al (2008) Resolution of liver cirrhosis using vitamin A-coupled liposomes to deliver siRNA against a collagen-specific chaperone. Nat Biotechnol 26:431–442

9. Tam Y, Chen S, Cullis P (2013) Advances in lipid nanoparticles for siRNA delivery. Pharmaceutics 5:498–507

10. Tam YYC, Chen S, Zaifman J et al (2013) Small molecule ligands for enhanced intracellular delivery of lipid nanoparticle formulations of siRNA. Nanomedicine 9:665–674

11. Wilner SE, Wengerter B, Maier K et al (2012) An RNA alternative to human transferrin: a new tool for targeting human cells. Mol Ther Nucleic Acids 1, e21

12. Ni X, Zhang Y, Ribas J et al (2011) Prostate-targeted radiosensitization via aptamer-shRNA chimeras in human tumor xenografts. J Clin Invest 121:2383–2390

13. Zhou J, Swiderski P, Li H et al (2009) Selection, characterization and application of new RNA HIV gp 120 aptamers for facile delivery of Dicer substrate siRNAs into HIV infected cells. Nucleic Acids Res 37:3094–3109

14. Dhar S, Gu FX, Langer R et al (2008) Targeted delivery of cisplatin to prostate cancer cells by aptamer functionalized Pt(IV) prodrug-PLGA-PEG nanoparticles. Proc Natl Acad Sci 105:17356–17361

15. Zhou J, Rossi JJ (2014) Cell-type-specific, aptamer-functionalized agents for targeted disease therapy. Mol Ther Nucleic Acids 3, e169

16. Reulen SWA, Brusselaars WWT, Langereis S et al (2007) Protein-liposome conjugates using cysteine-lipids and native chemical ligation. Bioconjug Chem 18:590–596

17. Leung AKK, Hafez IM, Baoukina S et al (2012) Lipid nanoparticles containing siRNA synthesized by microfluidic mixing exhibit an electron-dense nanostructured core. J Phys Chem C 116:18440–18450

18. Jayaraman M, Ansell SM, Mui BL (2012) Maximizing the potency of siRNA lipid nanoparticles for hepatic gene silencing in vivo. Angew Chem Int Ed 51(34):8529–8533

19. Jeffs LB, Palmer LR, Ambegia EG et al (2005) A scalable, extrusion-free method for efficient liposomal encapsulation of plasmid DNA. Pharm Res 22:362–372

Chapter 19

Screening of Genetic Switches Based on the Twister Ribozyme Motif

Michele Felletti, Benedikt Klauser, and Jörg S. Hartig

Abstract

The recent description of a new class of small endonucleolytic ribozymes termed twister opened new avenues into the development of artificial riboswitches, providing new tools for the development of artificial genetic circuits in bacteria. Here we present a method to develop new ligand-dependent riboswitches, employing the newly described catalytic motif as an expression platform in conjugation with naturally occurring or in vitro-selected aptameric domains. The twister motif is an outstandingly flexible tool for the development of highly active ribozyme-based riboswitches able to control gene expression in a ligand-dependent manner in *Escherichia coli*.

Key words Ribozyme, Riboswitch, Biosensor, Aptamer, Aptazyme

1 Introduction

The family of small endonucleolytic ribozymes is composed of RNA motifs that have a size in the range of 50–150 nucleotides and which are able of sequence-specific cleavage [1]. Until recently, only five classes of small endonucleolytic ribozymes were described: the hammerhead [2], the hairpin [3], the hepatitis delta virus (HDV) [4], the Varkud satellite (VS) [5], and the glmS ribozymes [6]. Lately, a new widespread class of small endonucleolytic motifs has been identified by Breaker and coworkers using a bioinformatics pipeline [7]. Breaker and coworkers coined the name "twister" because the secondary structure resembles the ancient hieroglyph "twisted flax." Twister ribozymes occur in both *Bacteria* (mostly in the class *Clostridia*) and *Eukarya* (fungi, plants, cnidarians, flatworms, insects, and fish). They are also found in a variety of environmental sequences (i.e., not lab-cultivated organisms). Although the physiological functions remain unknown, it was observed that the genetic contexts of twister ribozymes are often similar to those of the hammerhead ribozymes found in the same organism [7].

Günter Mayer (ed.), *Nucleic Acid Aptamers: Selection, Characterization, and Application*, Methods in Molecular Biology, vol. 1380, DOI 10.1007/978-1-4939-3197-2_19, © Springer Science+Business Media New York 2016

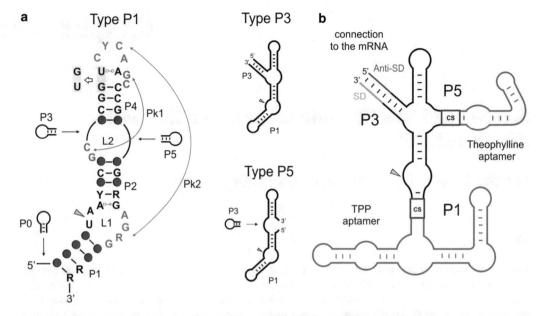

Fig. 1 The twister ribozyme as expression platform of artificial riboswitches. (**a**) Conserved consensus secondary structure of twister. The three twister types are represented with the possible structural variations. Conserved (75–97 %) nucleotides are shown. The positions in which the nucleotide identity is less conserved are represented by *red circles*. Pseudoknot 1 (Pk1) and pseudoknot 2 (Pk2) are represented in *green* and *blue*, respectively. The *arrowheads* identify the cleavage sites. The inactivating mutation is highlighted in *yellow*. (**b**) Example of an artificial two-input twister-based riboswitches. A type P3 twister was used as expression platform. The Shine-Dalgarno (SD) sequence of the reporter gene is inserted into stem P3 of the twister motif. Two different sensor domains are connected to the catalytic domain through suitable connection sequences (CS) in correspondence to positions P1 and P5

Twister ribozymes present a conserved consensus secondary structure which was first predicted by the group of Breaker and later confirmed and updated by three different groups trough crystallography (Fig. 1a) [8–10]. The conserved secondary structure includes three essential stems (P1, P2, P4) and eventually three optional stems (P0, P3, P5). Importantly the fold of the twister relies on the presence of two pseudoknots (Pk1, Pk2) whose topology results to be completely different from the one previously observed in the HDV and *glmS* ribozymes. According to the position of the termini, naturally occurring twister ribozymes are classified into three classes (Type P1, P3, P5).

Like the other small self-cleaving ribozymes, the twister motif also undergoes a sequence-specific internal transesterification reaction wherein the 2′-hydroxyl group of a ribose attacks the adjacent 3′-phosphate through an S_N2-like reaction mechanism. The cleavage products contain a 2′,3′-cyclic phosphate and a 5′-OH, respectively. The self-cleavage rate constant of a twister in a bimolecular construct under optimum conditions is estimated to be comparable to kinetics of the hammerhead ribozymes showing k_{obs} values around 1000 per minute. In particular the twister bimolecular construct described by Roth et al. exhibited a maximum cleavage rate at 1 mM Mg^{2+} and pH 7.4 [7]. The enhanced rate of catalysis of twister is due to (1) the orientation of the reactive groups for the in-line attack, (2) the neutralization of the negative charge on the non-bridging oxygen atoms of the cleavage site phosphate (transition-state stabilization), and (3) the deprotonation of the 2' oxygen nucleophile (general base catalysis). Until now no good hydrogen bond donor candidates involved in the neutralization of the developing negative charge on the 5' oxygen leaving group (general acid catalysis) have been identified [9].

The discovery of this new catalytic motif opens new frontiers in the field of RNA biology. In addition, it offers new tools for the construction of artificial genetic circuits to the synthetic biologists. In fact, taking advantage of the modular and hierarchical organization of many regulatory RNAs, ribozymes can be used in association with aptamers (generating the so-called aptazymes), to create artificial riboswitches able to control gene expression in vivo. In such an approach the ribozyme works as the expression platform of the riboswitch, whereas the aptamer domain makes the catalytic activity of the ribozyme dependent on the concentration of small molecules and metabolites [11–13]. The use of ribozymes as expression platform in artificial RNA switches reveals to be highly versatile with respect to the ligand specificity, the type of regulated RNA, as well as the host organism [14–20].

Twister ribozymes are formidable flexible tools for the development of artificial riboswitches for a variety of reasons. As aforesaid, twister ribozymes show to have an activity, at least in vitro, comparable to the one of hammerhead ribozymes and they are 100- to 500-fold faster than other small self-cleaving ribozymes [7]. Moreover, due to the fact that the twister ribozyme is found in both *Eukarya* and *Bacteria* [7], one can assume that this motif can be virtually employed for applications in different organisms. Finally, the twister presents multiple potential sites (P1, P3, P5) which can be modified in order to connect the catalytic domain to the messenger RNA or to aptameric sensor domains (Fig. 1b). The stem P1 is always present with a non-conserved sequence of variable length [7]. Although the nucleotides of the helix P1 are not directly involved in the formation of the active site, the presence of this structure is essential for the stability of catalytic core [8, 9]. The stem P3 is an optional structure that is located at the connection between one strand of Pk1 and the stem P4, whereas the stem P5 is sometimes found in correspondence of a loop, which connects stem P4 to stem P2 (Fig. 1a) [8, 9]. Both structures are accommodated in loops whose nucleotides are not conserved and which are not involved in the formation of the active site. Although the stems P3 and P5 are in general variable in the sequence and in the length, the published crystal structures suggest that their presence could have a critical role in the maintenance of the overall structure in some twister ribozymes [8]. The presence of multiple sites for the connection of aptameric sensor domains opens the way for developing riboswitches dependent on more than one ligand [21, 22].

Here we present a procedure for developing ligand-dependent twister ribozymes that are enable conditional control of gene expression in *E. coli*. This method is based on an aptazyme design strategy and an in vivo screening approach which have been developed in our group a few years ago to generate artificial hammerhead-based riboswitches [23]. In particular, one of the stems of the ribozyme moiety acts as a molecular scaffold for the sequestration of the ribosome-binding site (RBS) of an eGFP gene located on a suitable plasmid [23]. As an accessible Shine-Dalgarno (SD) sequence within the RBS is essential for efficient initiation of translation, the masking of the SD represent an easy and efficient way of controlling the expression of the mRNA of interest. Addition of ligands to the bacterial growth medium changes the activity of the ligand-dependent self-cleaving ribozyme, which in turn switches gene expression either on, if the RBS is released upon the cleavage, or off, if the cleavage is inhibited and the RBS is masked (Fig. 2). Ligand-dependent masking of the SD sequence is a common

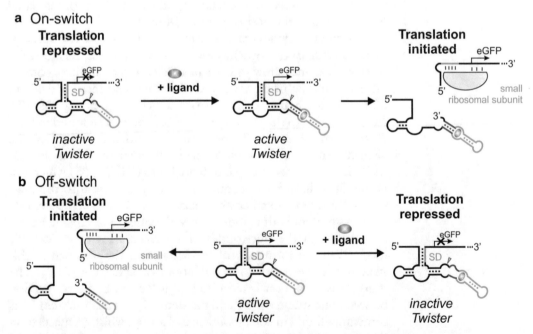

Fig. 2 Mechanisms of the twister-based on- and off-switches. In our approach the twister ribozyme acts as a molecular scaffold for the sequestration of the ribosome-binding site (RBS). Aptamer domains can be attached to the ribozyme as exchangeable ligand-sensing domains. Addition of ligands to the bacterial growth medium changes the activity of the ligand-dependent self-cleaving ribozyme which in turn switches gene expression on (**a**) or off (**b**). The use of eGFP as a reporter gene allowed a fast screening of clones by fluorescence

mechanism to control the expression of downstream genes in naturally occurring riboswitches. Using this strategy the twister can be used to switch on or off the expression of a reporter gene in vivo, combining them with natural occurring (e.g., TPP) or in vitro-selected aptameric domains (e.g., theophylline). The number of nucleotides in the connection sequence can be easily optimized in order to improve the switching activity.

2 Materials

All solutions are prepared using ultrapure water and reagents are of analytical grade quality. Reagents are stored at room temperature, if not otherwise noted.

2.1 PCR, Plasmid Purification, and Ligation

1. 0.5 M EDTA, pH 8.0: Dissolve 73.08 g EDTA and 30 g NaOH pellets in 400 mL H_2O, adjust pH 8.0 using NaOH, adjust volume to 500 mL.

2. 5× TBE buffer: Dissolve 54 g Trizma base, 27.5 g boric acid, 20 mL 0.5 M EDTA (pH 8.0) in 1 L of H_2O. Adjust to pH 8.3, if needed, with concentrated HCl. Dilute to 0.5× in H_2O before use.

3. 6× agarose gel loading buffer: Dissolve 4 mL 50 mM Tris–HCl, pH 7.6, 12 mL 100 % glycerol, 2.4 mL 0.5 M EDTA, 6 mg bromophenol blue, 6 mg xylene cyanol in 20 mL, prepare aliquots of 1 mL. Can be stored at –20 °C for several months.

4. Ready-to-use DNA size standard.

5. Agarose gels. For 0.8 % (w/v) dissolve 0.8 g agarose in 100 mL 0.5× TBE buffer by boiling in the microwave. Dissolved agarose can be stored in an incubator at ≥70 °C for several days. (*see* **Note 1**).

6. DNA oligonucleotides as primers to be used in PCR (*see* Subheading 3.1).

7. DpnI restriction enzyme and NEB buffer 4 (NEB).

8. Ethidium bromide gel staining solution. Dilute 200 µg in 400 mL H_2O. Can be kept in the dark at room temperature and reused (*see* **Note 2**).

9. Destaining solution: 400 mL H_2O. Can be kept at room temperature and reused.

10. PCR cycler.

11. PCR template: e.g., eGFP expression vector (pET16b_eGFP, AG Hartig, University of Konstanz).

12. Phusion Hot Start II High-Fidelity DNA polymerase, 5× HF buffer and 100 % DMSO (NEB). Store at –20 °C.

13. 2 mM dNTP Mix. Store at –20 °C.

14. Quick Ligase, 2× Quick Ligation Reaction Buffer (NEB). Store at –20 °C. Thaw Quick Ligation Reaction Buffer on ice.

15. DNA Gel Extraction Kit (e.g., Gel DNA Recovery Kit (Zymo Research)).

16. DNA Purification Kit (e.g., DNA Clean and Concentrator-5 (Zymo Research)).

17. Tabletop centrifuge.

18. Agarose gel electrophoresis chamber and power supply.

19. Razor blade.

20. UV light table.

2.2 Cell Culture and Screening

1. 96 deep-well plate and 96-well plate incubator (e.g., Heidolph Inkubator 1000 and Titramax 1000) (*see* **Note 3**).

2. Air-permeable adhesive seals for 96-well-plates.

3. 1000× carbenicillin stock solution (100 mg/mL): Dissolve 1 g carbenicillin in 10 mL 50 % (v/v) ethanol. Can be stored at –20 °C for several weeks.

4. Electro-competent *E. coli* (e.g., *E. coli* BL21 (DE3) gold, Stratagene) (*see* **Note 4**). Eighty microliter aliquots frozen at –80 °C.

5. Electroporator (e.g., Eppendorf Electroporator 2510) and electroporation cuvettes (e.g., Electroporation cuvettes, 0.1 cm (Bio-Rad)).

6. Fluorescence plate reader (e.g., TECAN M200).

7. LB-Carb-medium: For LB (Lennox) dissolve 10 g tryptone, 5 g yeast extract, 5 g NaCl in 1 L H_2O, adjust to pH 7.0, if necessary, and autoclave. Supplement with 1 mg/L carbenicillin before use. Can be stored at 4 °C for up to 4 weeks.

8. LB-Carb-agar plates: Dissolve 10 g tryptone, 5 g yeast extract, 5 g NaCl, and 10 g agar agar in 1 L H_2O and autoclave. Let cool until hand-warm, then supplement with 1 mg/L carbenicillin. Pour plates in 15 cm petri dishes. Plates can be stored at 4 °C for up to 6 weeks.

9. Kit for plasmid isolation (e.g., Zyppy Plasmid Miniprep Kit).

10. SOC medium: Dissolve 20 g tryptone, 5 g yeast extract, 0.6 g NaCl, 0.2 g KCl in 900 mL H_2O. Adjust pH to 6.8–7.0 using NaOH. Adjust volume to 960 mL and autoclave. Add 10 mL 1 M $MgCl_2$, 10 mL 1 M $MgSO_4$ and 20 mL 1 M glucose from sterile filtered stocks. Store 1 mL aliquots at –20 °C for several months.

11. Toothpicks sterilized by autoclaving.

12. Flat, transparent 96-well plates.

3   Methods

3.1   Construction of Randomized Riboswitch Libraries

Whole plasmid PCR is used to subclone a randomized plasmid library encoding eGFP under control of twister aptazymes. A pair of primers is designed to insert the aptamer sequence and a randomized connection sequence bridging the aptamer to the ribozyme (Fig. 3). Unbiased random nucleotides are generated during solid-phase DNA synthesis using a 1:1:1:1 mixture of nucleoside phosphoramidites. One primer needs to be 5′-phosphorylated for ligation, which is a prerequisite for T4 DNA ligase activity. We recommend the purification of primers that exceed a total length of 45 nt by denaturing PAGE (as described in [24]: Preparation of Denaturing Polyacrylamide Gels; Purification of Synthetic Oligonucleotides by Polyacrylamide Gel Electrophoresis).

1. Primers are designed according to following rules (see theophylline twister aptazyme example below):
   Theophylline twister aptazyme sequence:

   • Type P3 twister ribozyme connected to a theophylline aptamer in position P5 (underlined). The anti-Shine-Dalgarno and Shine-Dalgarno sequences are represented in the lower case.

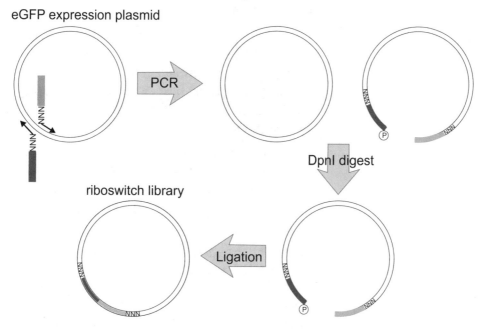

eGFP expression plasmid

PCR

DpnI digest

riboswitch library

Ligation

Fig. 3 Library construction: Whole-plasmid PCR is used for incorporation of new aptamer sequences using primers carrying randomized sequences that will constitute the connecting sequence. After PCR the template plasmid is degraded by DpnI digestion. The purified PCR product is ligated yielding a library of randomized riboswitches

Table 1
Protocol for setting up a PCR reaction

Starting concentration	Material	Volume	Final concentration
5×	HF buffer	10	1×
2 mM	dNTP mix	5	200 µM
100 µM	Forward primer	0.3	600 nM
100 µM	Reverse primer	0.3	600 nM
20 ng/µL	Template	0.5	0.2 ng/µL
100 % (v/v)	DMSO	1.5	3 % (v/v)
2 U/µL	Phusion Hot Start DNA polymerase	0.5	0.02 U/µL
	H₂O	31.9	
	Total	50	

- 5'cuccuuUAAAGCGGUUACAAGCCCGCAA **NN** <u>CA UACCAGCCGAAAGGCCCUUGGCAGG</u> **NN** AAU AGCAGAGUAAUGGGAAACCAUUAAUGCAGCU UUAaaggag 3'
  Forward primer:

- Introduce randomized nucleotides at the bridge between the aptamer (underlined) and the twister catalytic core

- Example: theophylline aptazyme

- 5' <u>GGCCCTTGGCAGG</u> **NN** AATAGCAGAGTAAT GGGAAACCATTAATGC3'
  Reverse primer:

- Introduce randomized nucleotides at the bridge between the aptamer (underlined) and the twister catalytic core

- Example: theophylline aptazyme

- 5' <u>TTTCGGCTGGTATG</u> **NN** TTGCGGGCTTGTAA CCGCT3'

2. Prepare the PCR reaction mixture as follows and split the mixture into three PCR tubes. The reaction can be scaled up, if necessary. We recommend using no more than 30 µL per tube (*see* Table 1).

3. Use the following conditions for the PCR with Phusion Hot Start II High-Fidelity DNA polymerase (*see* **Note 5**) (*see* Table 2).

4. Pool the PCR reaction and transfer into one 1.5 mL reaction tube.

Table 2
Description of thermal cycles during PCR

Step	Stage	Temperature (°C)	Time	Go to step	Repeat
1	Initial denaturation	98	30 s		
2	Denaturation	98	10 s		
3	Annealing	58	30 s		
4	Extension	72	20 s/kb	2	24
5	Final extension	72	7 min		
6	Cooling	4	Pause		

5. Purify PCR reaction mixture using DNA purification kit according to the manufacturer's recommendations. Elute DNA with 44 μL H_2O.

6. For removal of the template vector add 5 μL NEB buffer 4 and 1 μL DpnI. Mix thoroughly and incubate at 37 °C for 50 min. Heat inactive restriction endonuclease by 10-min incubation at 80 °C.

7. In the meantime, prepare a 0.8 % (w/v) TBE-agarose gel. A small pocket for the size standard and a large one for the PCR product is required. After DpnI digestion, add 10 μL of 6× agarose gel loading buffer and mix. Load the PCR sample and 2.5 μL DNA size standard. Run the gel at 10 V cm⁻¹ for 75–90 min.

8. After the gel electrophoresis, stain the gel into ethidium bromide staining solution. After 10 min transfer the gel into destaining solution for additional 10 min. UV light may cause mutations to DNA. To avoid exposure of the DNA sample to UV light, use the razor blade to excise the marker and a small part of the band with your sample. Use the UV light table for visualization of the DNA bands.

9. The correctly sized band is marked best by excision. Align the two gel fragments and excise the corresponding band in the non-irradiated gel fragment. Use the DNA gel extraction kit according to the manufacturer's protocol for purification. Elute DNA with 20 μL H_2O.

10. Measure DNA concentration based on A_{260} value of the sample.

11. Use the Quick Ligation Kit for the ligation of the PCR product. Mix 50 ng of the DNA sample in 9 μL of water with 10 μL 2× Quick Ligation Reaction buffer and adjust with H_2O to 19 μL. Add 1 μL Quick Ligase, and mix by pipetting. Incubate at 25 °C for 15 min.

12. Purify the DNA sample by using a DNA purification kit according to the manufacturer's instructions. Elute ligated DNA using 7 μL H_2O. Removal of the ligation buffer results in higher transformation efficiency.

3.2 Screening for Functional Aptazymes

The randomized Twister aptazyme library is incorporated into *E. coli* by transformation of the randomized plasmid. For library-scale transformation plasmids are transformed by electroporation of *E. coli* BL21 (DE3) gold ensuring high transformation efficiency.

1. Thaw 80 μL aliquot of electro-competent *E. coli* BL21 (DE3) gold on ice for 15 min.

2. Meanwhile, place one electroporation cuvette and an 1.5 μL aliquot of ligated DNA sample within a 1.5 mL reaction tube on ice.

3. Preheat 1 mL aliquot of SOC medium and LB-carbenicillin-agar plates at 37 °C.

4. Add 80 μL competent cells into the reaction tube containing the plasmid library and mix carefully. Transfer the mixture into the pre-chilled electroporation cuvette and transform by electroporation. Subsequently resuspend the cells in 1 mL pre-warmed SOC medium. Transfer the cells to an unused 1.5 mL reaction tube. Spread bacteria cells on pre-warmed LB-agar plates and incubate plates overnight at 37 °C (*see* **Note 6**).

Single colonies are individually investigated for switch performance in a high-throughput format. Single colonies are grown in microtiter plates and functional riboswitches are identified based on changes of the eGFP expression levels in the presence and absence of the aptamer ligand.

5. First, a bacterial clonal library is constructed in 96 deep-well, termed "master plates." Therefore, prepare 96 deep-well plates with 400 μL LB medium supplemented with carbenicillin are made. Allocate defined wells of the 96 deep-well plate for control cultures (*see* **Note 7**). Use sterilized toothpicks to pick single colonies from the agar plates. Remove toothpicks and seal plate with air-permeable adhesive cover. Incubate plates at 37 °C overnight in the 96-well plate incubator vigorously shaking.

6. Each master plate serves as template, which is used for final screening procedure. The screening of the clonal library requires the growth of each clonal culture once in the absence and otherwise in the presence of the aptazyme ligand. Therefore, prepare two additional destination plates. Add the medium into the wells so that medium without ligand alternates every column with medium supplemented with ligand (as shown in Fig. 4).

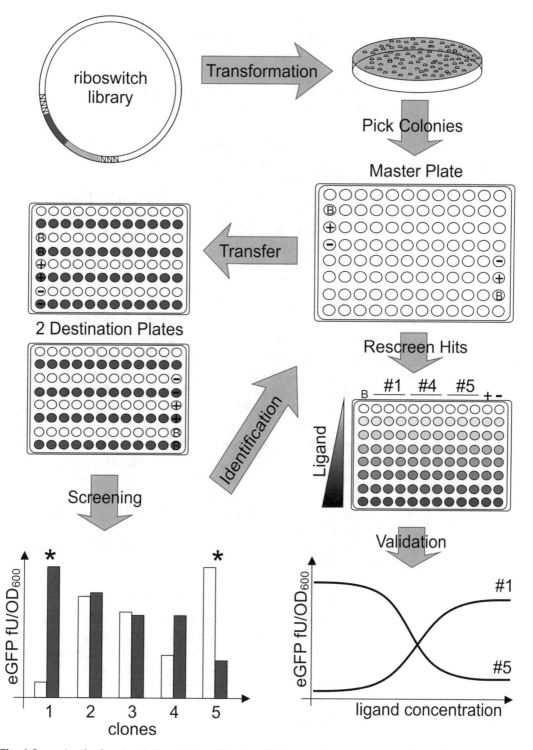

Fig. 4 Screening for functional riboswitches. The plasmid library of randomized riboswitches is transformed into *E. coli*. Individual clones are picked to construct clone libraries, which are screened for eGFP expression in the presence (*white*) or absence (*dark grey*) of the aptamer ligand. Controls for background fluorescence correction (B), positive control (+), and negative control (−) are included in the screening (*see* **Note 7**). Identified hits are further characterized

7. Use bacterial cultures of the master plates for the inoculation of cultures of the destination plates. Make sure that each clone is examined in the presence and absence of the ligand (*see* **Note 8**). Seal destination plates with air-permeable adhesive cover. Incubate the destination plates in the 96-well plate incubator at 37 °C overnight vigorously shaking.

8. Transfer 100 µL of the exponential phase culture into a 96-well microtiter plate and measure expression levels by determining eGFP fluorescence (excitation wavelength $\lambda_{ex} = 488$ nm, emission wavelength $\lambda_{em} = 535$ nm).

9. For analysis and the identification of potential genetic switches first subtract the background fluorescence from each fluorescence value. In addition measure OD_{600} for OD correction. Identify potential hits by comparing the fluorescence of each screened clone in the presence and absence of the ligand.

10. Clones of interest are inoculated in 5 mL LB medium and grown vigorously shaking at 37 °C for 8 h. Grown-out cultures are used for the preparation of glycerol stocks and the isolation of the plasmid encoding the aptazyme switch

11. Plasmids are isolated using a plasmid isolation kit according to the manufacturer's instructions. Sequence plasmids. Perform a sequence analysis to examine the integrity of the artificial riboswitch constructs and the identity of the randomized connecting sequence.

12. For further analysis investigate the riboswitches at increasing concentrations of the ligand. Prepare 96-well plates as detailed in **step 4** using a gradient of increasing ligand concentrations. Conduct experiments as triplicates. Plates are incubated shaking at 37 °C for greater 16 h before measuring eGFP expression.

Pool coverage is calculated by determining the oversampling factor (O_f) which is calculated as

$$O_f = T / V = -\ln(1 - P_i)$$

where T is the number of clones actually screened, V the maximum number of different clones of a randomized sequence (sequence space) and Pi the probability that a particular sequence occurs in the library [25]. A three-time oversampling of the theoretical sequence space results in a pool coverage of approximately 95 %.

---

4   Notes

1. Separation of 5–10 kb PCR products is best performed with 0.8 % (w/v) agarose gels. Higher percentages of agarose (e.g. 1.5 % (w/v)) result in better resolution for smaller products.

2. Ethidium bromide is toxic and mutagenic. Handle ethidium bromide according to the material safety data sheet (MSDS) and wear nitrile gloves while working with it.

3. For increased library sizes we recommend the use of 384 deep-well plates for cell culture and fluorescence measurements.

4. For the generation of electro-competent *E. coli* a detailed protocol can be found at www.eppendorf.com.

5. There are other polymerases as alternative to Phusion Hot Start polymerase that can be used. Adapt the PCR protocol and apply appropriate buffer and conditions as recommended by the manufacturer's protocol. We recommend the usage of a high fidelity polymerase. Primers should be designed to exhibit an annealing temperature of 60–63 °C. For improved PCR amplification we suggest to determine the optimal annealing temperature by conducting a gradient PCR.

6. The number of colonies depends on the transformation efficiency of the electro-competent cells used and the volume that is plated. Cross-contamination needs to be avoided during the screening process. To yield single colonies plate dilutions of the transformed cells.

7. Always include a positive control, a negative control, and a culture for subtraction of the background fluorescence (*E. coli* transformed with a plasmid not containing the eGFP gene). As positive controls use cells that are transformed with a plasmid expressing eGFP either under control of a catalytically active twister ribozyme or without ribozyme insert. As a negative control transform a plasmid that features a catalytically inactive Twister ribozyme variant (mutation highlighted in yellow in Fig. 1a), unable to free the ribosomal binding site. For determining the background fluorescence is transformed with a plasmid that is not encoding eGFP (e.g., pET16b). Note that all plasmids should transfer the same antibiotic resistance marker. As one master plate will be distributed to two destination plates, we recommend the inclusion of the controls on the top half and on bottom half of each master plate. Thereby, each destination plate will receive a set of controls. We also recommend not using the wells in the corners for screening purposes as growth of the cells will suffer from evaporation of the medium.

8. Minimal medium enables improved control over medium composition. Therefore, when screening for aptazymes that are triggered by natural ligands (e.g., TPP), we recommend the usage of minimal medium, (e.g., M9 or M63 medium). In addition, consider the membrane permeability of you ligand. Not all ligands are readily taken up into the cells; e.g., the screening for TPP riboswitches required the supplementation

with thiamine to the medium, which is processed by the intra-cellular thiamine kinase and thiamine phosphate kinase to yield TPP. In addition, other ligands might not be stable in the medium. Before applying new ligands or media we recommend measuring growth curves for the bacteria, to rule out any potential toxic effects of the ligand.

References

1. Ferre-D'Amare AR, Scott WG (2010) Small self-cleaving ribozymes. Cold Spring Harb Perspect Biol 2(10):a003574

2. Prody GA, Bakos JT, Buzayan JM, Schneider IR, Bruening G (1986) Autolytic processing of dimeric plant virus satellite RNA. Science 231(4745):1577–1580

3. Buzayan JM, Hampel A, Bruening G (1986) Nucleotide sequence and newly formed phosphodiester bond of spontaneously ligated satellite tobacco ringspot virus RNA. Nucleic Acids Res 14(24):9729–9743

4. Sharmeen L, Kuo MY, Dinter-Gottlieb G, Taylor J (1988) Antigenomic RNA of human hepatitis delta virus can undergo self-cleavage. J Virol 62(8):2674–2679

5. Saville BJ, Collins RA (1990) A site-specific self-cleavage reaction performed by a novel RNA in Neurospora mitochondria. Cell 61(4):685–696

6. Winkler WC, Nahvi A, Roth A, Collins JA, Breaker RR (2004) Control of gene expression by a natural metabolite-responsive ribozyme. Nature 428(6980):281–286. doi:10.1038/nature02362 nature02362 (pii)

7. Roth A, Weinberg Z, Chen AG, Kim PB, Ames TD, Breaker RR (2014) A widespread self-cleaving ribozyme class is revealed by bioinformatics. Nat Chem Biol 10(1):56–60

8. Liu Y, Wilson TJ, McPhee SA, Lilley DM (2014) Crystal structure and mechanistic investigation of the twister ribozyme. Nat Chem Biol 10:739–744

9. Eiler D, Wang J, Steitz TA (2014) Structural basis for the fast self-cleavage reaction catalyzed by the twister ribozyme. Proc Natl Acad Sci U S A 111(36):13028–13033

10. Ren A, Kosutic M, Rajashankar KR, Frener M, Santner T, Westhof E, Micura R, Patel DJ (2014) In-line alignment and Mg2+ coordination at the cleavage site of the env22 twister ribozyme. Nat Commun 5:5534

11. Vinkenborg JL, Karnowski N, Famulok M (2011) Aptamers for allosteric regulation. Nat Chem Biol 7(8):519–527

12. Frommer J, Appel B, Muller S (2015) Ribozymes that can be regulated by external stimuli. Curr Opin Biotechnol 31:35–41

13. Berens C, Groher F, Suess B (2015) RNA aptamers as genetic control devices: the potential of riboswitches as synthetic elements for regulating gene expression. Biotechnol J 10(2):246–257. doi:10.1002/biot.201300498

14. Klauser B, Atanasov J, Siewert LK, Hartig JS (2014) Ribozyme-based aminoglycoside switches of gene expression engineered by genetic selection in S. cerevisiae. ACS Synth Biol 4(5):516–525

15. Wieland M, Benz A, Klauser B, Hartig JS (2009) Artificial ribozyme switches containing natural riboswitch aptamer domains. Angew Chem Int Ed Engl 48(15):2715–2718. doi:10.1002/anie.200805311

16. Wieland M, Auslander D, Fussenegger M (2012) Engineering of ribozyme-based riboswitches for mammalian cells. Methods 56(3):351–357

17. Gu H, Furukawa K, Breaker RR (2012) Engineered allosteric ribozymes that sense the bacterial second messenger cyclic diguanosyl 5′-monophosphate. Anal Chem 84(11):4935–4941. doi:10.1021/ac300415k

18. Klauser B, Hartig JS (2013) An engineered small RNA-mediated genetic switch based on a ribozyme expression platform. Nucleic Acids Res 41(10):5542–5552

19. Nomura Y, Zhou L, Miu A, Yokobayashi Y (2013) Controlling mammalian gene expression by allosteric hepatitis delta virus ribozymes. ACS Synth Biol 2(12):684–689. doi:10.1021/sb400037a

20. Saragliadis A, Hartig JS (2013) Ribozyme-based transfer RNA switches for post-transcriptional control of amino acid identity in protein synthesis. J Am Chem Soc 135(22):8222–8226. doi:10.1021/ja311107p

21. Klauser B, Saragliadis A, Auslander S, Wieland M, Berthold MR, Hartig JS (2012) Post-transcriptional Boolean computation by

combining aptazymes controlling mRNA translation initiation and tRNA activation. Mol Biosyst 8(9):2242–2248. doi:10.1039/c2mb25091h

22. Win MN, Smolke CD (2008) Higher-order cellular information processing with synthetic RNA devices. Science 322(5900):456–460

23. Wieland M, Hartig JS (2008) Improved aptazyme design and in vivo screening enable riboswitching in bacteria. Angew Chem Int Ed Engl 47(14):2604–2607. doi:10.1002/anie.200703700

24. Sambrook J, Russel DW (2001) Molecular cloning: a laboratory manual, 3rd edn. Cold Spring Harbor Laboratory, Cold Spring Harbor, NY

25. Reetz MT, Kahakeaw D, Lohmer R (2008) Addressing the numbers problem in directed evolution. Chembiochem 9(11):1797–1804. doi:10.1002/cbic.200800298

INDEX

Günter Mayer (ed.), *Nucleic Acid Aptamers: Selection, Characterization, and Application*, Methods in Molecular Biology, vol. 1380, DOI 10.1007/978-1-4939-3197-2, © Springer Science+Business Media New York 2016

Printed in the United States
By Bookmasters